Imagine
becoming Earth's
greatest stewards.

LIEN DE RUYCK

OFF-GRID
Adventures

Lannoo

20 OFF-GRID
LOCATIONS

The numbers on the map refer to the chapter numbers.

LIVING ADVENTUROUSLY

A commitment to change and curiosity

Exploring the world means something different today than it did, say, 100 years ago when Ernest Shackleton ventured into the icy South Atlantic Ocean in a sailing boat. Now we have over 4,500 satellites swarming around us, their zoom capabilities transporting us to every far-flung corner of the world. We're launching tourists into space, running marathons in Antarctica and getting stuck in human traffic on Mount Everest. Is there anything left to explore? What does it mean to live adventurously in the 21st century? And how could we become a new kind of explorer in the future?

For me, off-grid adventures are about being able to 'disconnect'. Not necessarily from running water or the grid, but from the obvious, somewhat ingrained path. Being able to create a bit of distance, without a productivity meter, watch or plan. Get a bird's-eye view. Rise above your own life. With that openness and wonder that we often leave behind in our childhood.

This book has become a collection of the places and encounters that have left the biggest impression on me. Along a breadcrumb trail from Canada to Antarctica, from Rwanda and Japan to Kyrgyzstan, it gives you insights into landscapes, cabins, living rooms and other ways of life. Come along for the ride, on foot or horseback, on a motorbike or in the hull of a ship. Because experiences are so much more valuable if you can share them with others: the anticipation, the beauty, the doubts, the confrontation. Told without any filter, and with as much appreciation for the beautiful as the slightly rough-around-the-edges parts.

More than ever, we need adventurers who want to show strength, guts, determination and courage.

Old-school adventure

I used to devour them, the stories of the physical and mental struggles of the adventurers who risked their lives to do the seemingly impossible. It was a dream, like it was for them, to go where no one had gone before. I admired their courage, resilience and adaptability. Skills that gave us hope and showed us what we're capable of as human beings. I wanted that too. Until the other side to that romantic story also became more and more visible.

European explorers who mapped their 'terra novas' were hailed as heroes, but carnage often followed in their wake. No matter how heroic their logbooks, most knew exactly what would happen to the native population and the animals they recorded. Also the stories of their expeditions and exploits rarely featured the achievements of women or local helpers. Those footnotes started to grate more and more. With freedom comes responsibility.

From conquerors

Today the lure of adventure seems to be everywhere. The logbook has become a blog, the national flag a selfie. Away from our harness of comfort, we're always looking for something wilder, something more authentic. A new service industry emerged, but at what cost? In a photo taken at the top of Mount Everest in May 2019, a long queue of climbers is waiting to reach the crowded peak during a short window of good weather. Such hubris that would ultimately result in the death of half a dozen climbers, also risking the lives of their Sherpas. A climb to be able to win and overcompensate. The ego trip, driven to the top.

What do we want to prove to ourselves? Why do we even want to climb that mountain? 'Because it's there' no longer seems like a good enough answer. We can no longer argue that it's for the greater good: for example, cartography, anthropology or botany. Now that much of our Earth's surface has been explored, the challenges are more personal. Even if they are for a good cause.

To regenerators

Yet we can still use adventurers. More than ever, we need adventurers who want to show strength, guts, determination and courage. Because there are still enormous challenges, now social and environmental, that we need to overcome. The next wave of explorers could regenerate all that beauty: protect it, restore it and help it flourish. This is how together we can once again do something that no one has ever done before. A journey of discovery towards the 22nd century, a journey that is subservient and nature-boosting, instead of controlling and dominating. 'It's not the mountain we conquer, but ourselves' takes on an even more beautiful meaning. Even ordinary people can do extraordinary things.

My best life lesson so far? That acidification only happens if you don't train certain muscles enough or if you give them the wrong fuel. So train that open mind and nourish yourself with positivity. Be your own satellite, and occasionally travel to your extremes. Deviate. Roam free. Fall apart. Rearrange all your atoms. Be mild. Stay put. And surround yourself with those who believe in the goodness of people. Because the future is also joyful, exciting and hopeful. We become what we believe. And that adventure starts here, now. //

Lien De Ruyck | @sonderling.be

CURIOUS PLACES

A Quiet Zone

Listening to the whispers of the universe

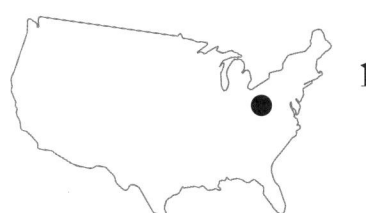

1

Did you know there's a village where, for 50 years, time seems to have stood still? Where no smartphones glow, and microwaves are banned? Well hidden and protected by the highest peaks of the Snowshoe Mountain ski resort in West Virginia is the Green Bank Telescope, one of the biggest listening devices on Earth. It sits in the middle of a gigantic quiet zone, where scientists listen day and night to the soundtrack of the universe.

Ever since I decided to travel across the US, there's been one place on my list: the National Radio Quiet Zone (NRQZ). Deep in the forests of West Virginia, the state that's two-thirds national park, you'll find an officially designated quiet zone. This unique area was not created so the locals, animals or yogis can get a good night's sleep, but in the name of science – and the search for answers to questions about space and extraterrestrial life, hundreds of light years away.

Phone home

The seven telescopes and hundreds of instruments at the Green Bank Observatory detect radio waves from space. The Green Bank Telescope is the largest steerable dish radio telescope – and the third largest manoeuvrable object – in the world. With its 100 metres (328 feet) diameter, it is the most sensitive receiver on Earth. Wider than two football fields, higher than the Statue of Liberty and taller than the world's highest roller coaster. The observatory is a hotspot for scientific searches for extra-terrestrial intelligence (SETI), but the down-to-earth locals simply call it 'The Very Big Thing'.

Dotted around the zone are several radio dishes, including the 43 metres (141 feet) telescope protruding through the misty fields here.

Part of the Green Bank Observatory, which dates from the 50s, the Green Bank Telescope cost no less than $95 million and achieved first light in 2000, going into operation in 2003. The most accurate, versatile, large-dish radio telescope in the world, it is super-sensitive to the faint clouds of hydrogen that hang out between the stars and galaxies. Scientist using the telescope have discovered a two-solar pulsar and a huge 'superbubble' of hydrogen gas rising nearly 10,000 light-years above our Milky Way galaxy.

But this came at another high price: to avoid interfering with the reception of the cosmic radio waves, a 13,000 square-mile (34,000km²) quiet zone was set up in 1958 around the observatory. Within that zone, there are a whole lot of restrictions on technology: Wi-Fi is banned, as are other mod cons like mobile phones, microwaves and electrical appliances. All in the name of science, the village of Green Bank went completely dark.

Off the grid

When we drive into the zone, we lose our mobile phone signal and our Bluetooth gets automatically disconnected. 'You are now Entering the West Virginia Radio Quiet Zone', announces a roadside sign. Once we're in the inner circle, where the strictest rules apply, no one from the outside world can reach us for 16 kilometres (10 miles). No modern technology is allowed here. Except of course in the observatory's control room, which stands inside a Faraday cage. While the astronomers use their high-tech tools to stare at dead stars, far-flung galaxies and signs of extraterrestrial life... the world around them seems to have come to a standstill.

Here, electromagnetic silence is sacred.

If you're now picturing a peaceful retreat, then think again. Because the place seems to have a strange appeal for eccentric figures from all walks of life. Its 180 residents are a motley crew of long-time locals, bear hunters and astrophysicists as well as charlatans, cult leaders and even – we are told – murderers. Some survivalists are here getting ready for the kind of mass power outage that would turn the entire planet into a quiet zone. A hippie commune is making a stand against the 'death culture' of the outside world. A group of neo-Nazis have withdrawn to their own military base in the hills. There's even a community of around a hundred sufferers from the controversial electromagnetic hypersensitivity. Each of them is trying to build a new life here, safely out of mobile phone and Wi-Fi range.

Back to basics

Here, electromagnetic silence is sacred. So residents and employees make do with diesel cars and appliances from the 60s and 70s, because even a petrol car's ignition system causes too much interference. Locals have had to come up with creative ways to communicate. Anyone who deliberately breaks the rules can expect an immediate visit from an investigation team. The observatory's electromagnetic interference tracking truck, or EMITT, detects all illegal radio waves. If you want to use a new device, you first have to get it

A view over the edge of the Green Bank Telescope's main dish. *(right)*
The Green Bank Telescope, also called 'The Very Big Thing' by locals. *(following pages)*

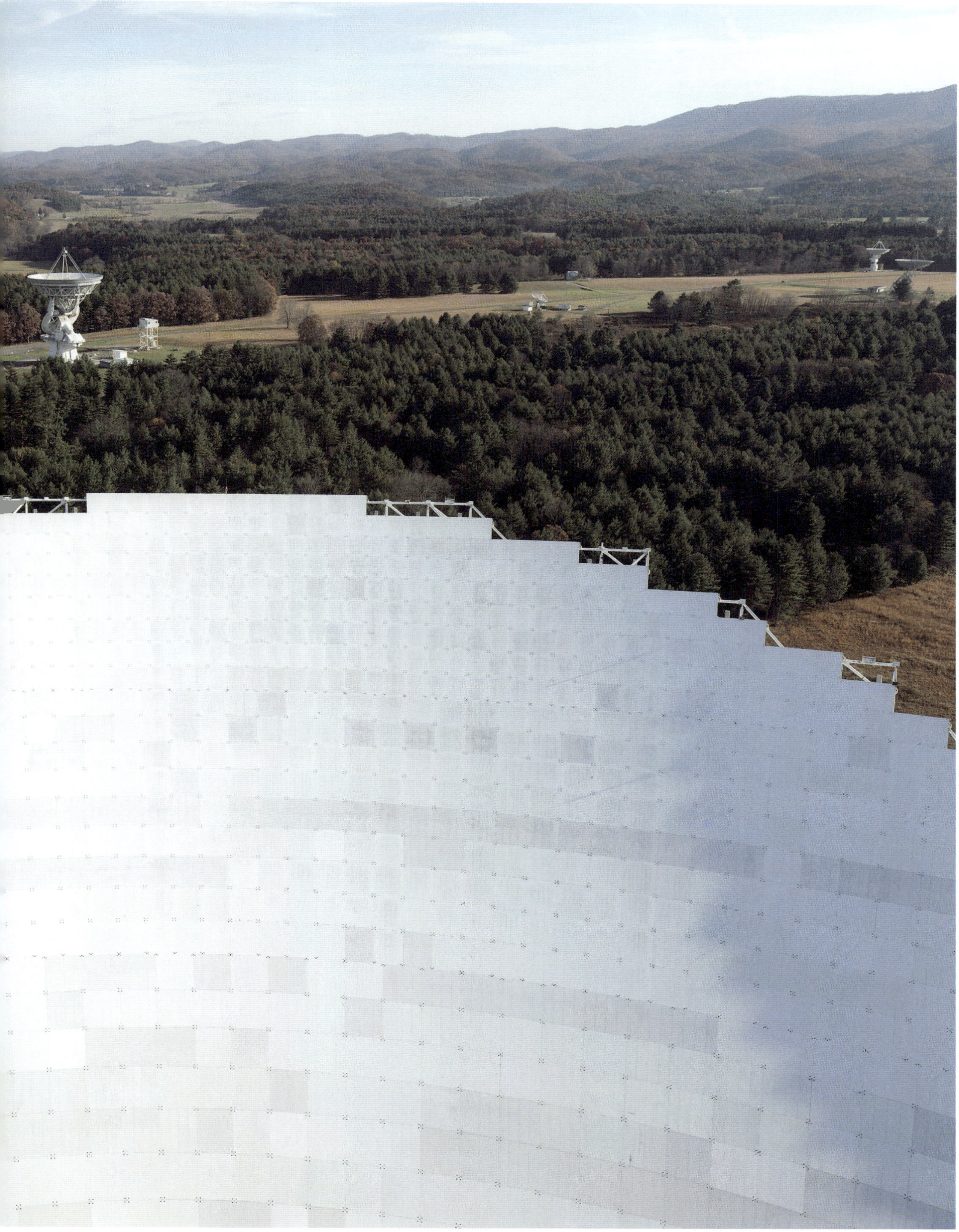

'For the electrosensitives seeking relief from their pain, for the astronomers in need of a quiet sky, for the hippies desiring a peaceful landscape, for the tech-addicted tourists forced to go offline, the Quiet Zone was an unexpected refuge. It was an escape, at its best, from ourselves.'

Stephen Kurczy, *The Quiet Zone*

With its restrictions on technology, the area around Green Bank attracts people who want to live off-grid. Their electrical appliances resemble artefacts from another lifetime.

Green Bank Observatory
West Virginia, USA

CURIOSITY

Low-tech 'quiet zone'. Homeground of the Breakthrough Listen project, which has searched the closest 1,327 stars in our galaxy for hints of intelligent life and advanced civilizations.

ADVENTURE TIME

Best explored in 1-2 days. Chances of hearing alien civilization conversations are estimated to be 1 in 400,000 years, whatever the season.

TRAVEL TIP

Pre-register for a 'Radio Astronomer for a Day' overnight educational trip in the science centre, or apply for volunteer work on the site.

BEFORE YOU GO

Read *The Drake Equation* by Paul Kranzler and Andrew Phelps, who spent many weeks in and around the small town of Green Bank, documenting this unique community.

approved. So don't even think about sneaking in a selfie here. The telescope is so sensitive that, even if you took a selfie on Saturn in aeroplane mode, it would still pick up the signal. But sometimes things go wrong. The SETI research once ground to a halt for three months after some flying squirrels with GPS trackers infiltrated the NRQZ. *Eat this, aliens.*

Maintaining this silence is nothing short of an astronomical task. Not only due to the decrease in government subsidies, but also to the increase in terrestrial radiation and space activities. To really understand it, you have to experience first-hand this interaction of the deep cosmos, the high-tech world of astronomy and the subcultures in the village. Ready to take your digital detox to the next level? //

A behind-the-scenes look at one of the old telescopes, now used for educational purposes.

A Hope Spot

Freediving in the Indian Ocean

2

Mozambique invites exploration at a slow pace. We're headed to the coral island of Vamizi, set among the Quirimbas Archipelago. This eight-mile-long finger of island is one of the world's most important biodiversity hotspots. A bastion of hope, it offers a deep dive in paradise. To get there, we admire Africa from above, flying low over the clearest blue water, with tiny remote islands all around us. Easing down, until our wheels touch the island airstrip and roll us to a gentle stop. Is it me, or does time seem to have slowed down already?

The island is blessed with a tropical climate and powder-soft sandy beaches that rival those of the Maldives. We stay in one of only six secluded villas on the island. Nestled in the shadow of mature casuarina trees, each has its own access to the beach. An easy encounter with the untamed, so far. I can't wait to explore and jump into a kayak to get a first glimpse of the environment. As it's still early morning, kingfishers and cicadas fill the air with their dawn songs. I glide almost effortlessly over the glass like water and drift past mangroves along the shoreline. With tropical temperatures of about 35 °C, I'm looking forward to spending time in the water and exploring the healthy coral reefs around.

Butterfly effect

All over the tropics, reefs are in terminal decline, prey to global warming, overfishing and pollution. In parts of the Indian Ocean, bleaching has affected coral by 60-90%. But we live in hope: in 2013, this island group was designated as one of the first Hope Spots by Mission

The art of freediving, a serene underwater dance.

Blue, an alliance of 200 respected ocean conservation groups and like-minded organizations. Hope Spots are seen as parts of the world that are critical to its health and its ecosystem. Any successful result here has an exponential knock-on effect for the world as a whole. If you want to see the butterfly effect at work, this is a good place to start.

But change doesn't just happen overnight. After years of effort, the community-led Friends of Vamizi programme has managed to establish a dedicated research centre with resident zoologists on the island. Together they have recorded over 180 different species of coral and more than 300 species of reef fish. This is a sanctuary for some of the most significant wildlife habitats in the Indian Ocean and a community-driven marine protected area (MPA), in partnership with the WWF and UNESCO.

If you're lucky enough to visit the island in July or September, you can spot the humpback whales on their epic voyages between East Africa and Antarctica. They bring their newborn calves with them and the new families enjoy frolicking in the deep water channels around the island.

Underwater world

No wonder that this is the place divers dream about… The wild underwater landscapes, the bounty of tropical ocean currents and an abundance of marine life make Vamizi diving beyond world-class. Neptune's Arm – where coral gardens tumble down the edge of a 1,000-metre (3,280-foot) cliff so crowded with fish you can hardly see the view – has been named as one of the top ten scuba sites on the planet.

Any successful result here has an exponential knock-on effect for the world as a whole.

I grab a snorkel, mask and set of fins, while my companions are planning a dive into the deep blue. The best spots are reached by motorboat with a personal guide, as few of the diving sites have been properly mapped yet. We are headed to an underwater plateau where a coral reef plummets into a deep ocean channel, while a pod of dolphins swim alongside shyly. After a half-hour trip the line of the darkest blue water marks the drop-off. The scuba divers make their backflip into the water and disappear in no time into the seabed 500 metres (1,640 feet) below. I ditch the scuba tank, eager to savour this raw natural beauty with no equipment.

The snorkelling instructor introduces the rest of us to the art of freediving, while dancing with the drift. We float on the surface and take slow, deep breaths to slow down our pulses. Diving with nothing but a mask and some fins is your best shot to blend in with the environment. Professional freedivers stay underwater for more than 11 minutes, in a serene pursuit of feeling at one with the ocean itself. The no-limits freedive record is held by Herbert Nitsch. He descended 253.2 metres (830 feet), which earned him the title of 'deepest man on earth'. No pressure, right?

As my yoga teacher at home is still trying to teach me how to breathe properly ashore, I go for a baby freedive. The underwater landscape is breathtaking at first, with yellow, green, brown and black sponges. Some look like spilled paint, others grow like fingers sticking out from the wall and some are barrel corals with skeletons as wide as I am tall. A cradle of coral for the whole region, this mother reef has a unique chance of being resilient in the face of climate change. A place where the ocean and humankind still thrive side by side. Where scientists and locals fight to stop the damage before it starts.

Her Deepness

I spot a giant seashell as a tiny school of clownfish passes by. I feel surrounded by life. In a blink, a wave of emotions washes over me. I feel so privileged to swim in these healthy waters, which we have taken for granted for so long. While looking up towards the shining surface of the sea, scraps of a conversation with Sylvia Earle, oceanographer and founder of Blue Mission, pop into my mind: 'The ocean is not just water, rocks, sand, whatever. The ocean is a living system, every spoonful that you look at. We think of life in the sea in terms of fish and whales and coral reefs and the like, but most of the action is very small, microscopic and submicroscopic.' In 1979 Earle became the first person to walk solo on the bottom of the sea, under a quarter mile of water. Her fellow scientists have referred to her as 'Her Deepness' ever since. She led a historic team of all-female aquanauts, living for two weeks in an enclosed habitat on the ocean floor. While the whole world looked up, fascinated with outer space, she sank down in 'inner space', discovering alien worlds beneath Earth's waters.

We stay in the water for about an hour, going up and down again. Taking in the silence beneath the waves. Learning how to clear our snorkels, while equalizing pressure under water. This all goes rather smoothly, until my heart skips a few beats when a huge tuna passes by. Mistaking it for a white shark, as both have stiff bodies and tails that allow them to swim in bursts, I accidentally swallow a big gulp of seawater. The diving instructor is so amused that he too has to go up to catch a breath.

Coconut crabs

On the boat trip back, one of the divers shares his close encounter and Eskimo kiss with a giant green turtle. On Vamizi island, adventure is not only to be found underwater. There is always something extraordinary to see in this Eden; the island itself is home

Sunset on a traditional dhow. *(previous pages)* **Surrounded by a swirl of bubbles after a school of fish has passed by.** *(right)*

to the largest recorded population of green turtles in Mozambique, managed by one of the longest standing turtle monitoring programmes in East Africa.

The island is also the natural habitat of the coconut crab, the largest terrestrial arthropod in the world. This striking creature climbs trees, cracks coconuts open with its massive claws, and is known as 'the robber crab' because it raids birds' nests. Legend has it that coconut crabs even made part of Amelia Earhart's body disappear, when her plane crashed somewhere over the southern Pacific in 1937. Two years after her disappearance, scattered skeletal remains were found on the desert island of Nikumaroro, where these crabs reign. An unsolved whodunnit until today. When they come

out at dusk to look for food, you'd rather keep your shoes on, to say the least.

We see the day out on a traditional dhow, with perfect sailing conditions. Their billowing white sails move silently against the blue expanse of the Indian Ocean. Dhows were traditionally used as trading vessels, to carry heavy items such as fruit, fresh water or merchandise, along the coasts of Eastern Arabia, East Africa, Yemen and coastal South Asia. Nothing afloat is so graceful as these simple wooden fishing boats. //

Island hopping around Vamizi, snorkelling (*left*) **and from the plane** (*right*) **as it glides right over the ocean.**

'Oceans truly have no borders.
They are the ultimate commons, connected
by the salty waters that flow around the globe –
a fragile blue expanse that connects us all,
and on which we are all intimately dependent.'

Sylvia Earle

LOCATION

Vamizi Island
Mozambique, Africa

CURIOSITY

Its crown jewel is Neptune's arm, a reef wall plunging
dramatically to below 200 metres (650 feet). A breeding zone
for grey reef sharks, reef fish species and turtles.

ADVENTURE TIME

Best explored in 4-5 days. Whale encounters in July or
September.

TRAVEL TIP

The island is only accessible by charter flight to Vamizi airstrip
from Dar es Salaam international airport. Visit the turtle
conservation centre (the nesting season is in March and June).

BEFORE YOU GO

Watch the award-winning documentary Cradle of Coral by
WWF ambassador Mattias Klum about this area defined as a
Hope Spot by the Mission Blue foundation.

**Coral reef fish going into
hiding between the yellowy
brown skeleton of the fire
coral. *(right)* Face to face
with a school of curious
horse mackerel. *(following
pages)***

3

A Twilight Zone

Shipwrecks and stargazing on Terschelling

The Wadden Island of Terschelling awaits travellers just two hours from the Dutch mainland. But only visitors who have the patience of a saint should make the journey, because as soon as you set sail for the island, it's the tide that sets the pace.

The Wadden Islands are a treasure trove, and not just for their nature. During storms here, hundreds of ships were smashed to smithereens and washed ashore as wreckage. The stories are there for the taking on this strip of land whose inhabitants once collaborated with pirates and soldiers. Terschelling has always been a haven for seabirds, seals and other marine life, but also for the people from all around the world who have washed up on its shores. Today, place names such as the Dark Forest, Corpse Road and Dead Man's Coffin leave little to the imagination. At night, the island turns into one of the darkest places on Earth, even revealing the Milky Way. A shady world worth exploring from top to bottom.

We reach the island after a two-hour crossing. Captain Jan Zorgdrager and his crew know the secrets and pitfalls of the mudflats and their sandbanks better than anyone. Invisible forces are at work on Terschelling. Because of the erratic tide, the weather, wind and waves, you never know exactly what to expect. So no two days are the same on the bridge of the *MS Friesland*. Radar, VHF radio, depth gauge and many trained sailors' eyes guide us in a zigzag motion 30 kilometres (19 miles) out from the rift. Always beautiful, always different.

Twilight view of the Drenkelingenhuisje, a former shelter for castaways.

Terschelling is officially one of the largest municipalities in the Netherlands, but the island itself is only 5 kilometres (3 miles) wide on average. Only 87 km² (33 square miles) of its 712 km² (275 square miles) are land, with the mudflats appearing and disappearing around it.

A magical place brimming with life, the Wadden Sea is the largest unbroken system of intertidal sand and mud flats in the world, and was declared a UNESCO World Heritage Site in 2009. In this dynamic landscape shaped by wind and tides, flora and fauna flourish. The Wadden Sea is home to more than 10,000 plants and animals, many of which are rare. A mooring place for seals, you can spot porpoises here, and each year ten million birds come to breed for the winter. It is an irreplaceable ecosystem, the responsibility for the conservation of which is shared between the Netherlands, Germany and Denmark, because biodiversity on a global scale depends on the Wadden Sea.

When I first heard the name Wadden Islands, I imagined somewhere warmer. But the islands are in the North Sea, north of the Netherlands and Germany, and west of Denmark. The main islands are Texel, Vlieland, Terschelling, Ameland and Schiermonnikoog, while nine smaller islands are uninhabited and part of a designated nature reserve.

Bunkers, beacons & berries

Terschelling attracts those seeking not only silence but also fortune. Since the 15th century, many skippers have been caught off guard by the treacherous tidal channels and maze of sand trenches, explains Captain Zorgdrager as we dock. Thousands of treasures, from flotsam and wooden pirate legs to pigs and Nike trainers, have been washed ashore over the

Beachcombing has been elevated to a profession and a real lifestyle here.

centuries. Beachcombing has been elevated to a profession and a real lifestyle here. There's always something to pocket, although unfortunately these days it's likely to be plastic junk.

The largest and most precious finds are taken to the Wrakkenmuseum (Wrecks Museum) on Formerum. The museum is even built from the wreckage of a three-masted Cyprian barque, which sank in 1905. Inside, you'll find a collection of washed-up oddities. The wreck map shows more than 150 ships that have sunk off the island's coastline. Even today there's a lively dive scene here, the gold bars and silver pieces continuing to beckon through the thick layers of sand. But the enormous currents that rush through the tidal inlets between the islands at every tide complicate the search. What is visible today may be gone tomorrow.

The Drenkelingenhuisje also looks like it was washed up on the endless sandy plain. From the Heartbreak Hotel, the last beach pavilion on the island, I walk out to this abandoned little hut on stilts. Castaways who managed to haul themselves out of the churning surf and into the wooden shelter would then hoist the large open sphere that dangles in the wind into the air – a visual cry for help. Today it is mainly walkers who end up stranded at the Drenkelingenhuisje. After a trek of 15 kilometres (9 miles), they can quench their thirst with a Schoemrakker, a local amber beer that slides down easily with some kibbeling fish bites.

A little further along the beach I find traces of the Second World War. Stories of horror, heroes and resistance fill the air as a battle once raged between the Allies and the German air force in the airspace above the island. Almost all bunkers along the island's coastal strip were part of the historic Atlantic Wall. Bombers

From the deck of the *MS Friesland*. (top left) Mudflat walking on the Boschplaat. (right) Gathering cockles. (bottom left) West-Terschelling pier. (bottom right)

were monitored from the hidden radar positions, which we discover while cycling deeper into the island's forests. The Tigerstelling is a bunker complex where the battle of the skies over the Wadden Islands has been made visible again in a concrete jungle. The towering structures have been restored to their original state and made accessible thanks to bunker paths, maps and treasure hunts for the little ones.

Yet the most famous washed-up islander is also the smallest: the cranberry. The story goes something like this. About a hundred years ago, a beachcomber on Terschelling found a barrel washed ashore, which must have been lost by an American ship during a storm. To his regret, it wasn't filled with expensive brandy, but with sour, hard, red berries. These were kept on board American ships as a cure for the sailor's disease, scurvy. Disappointed, he threw the cranberries into the dunes. Little did he realize that years later the berries would cover the dune valleys. I struggle to escape it in the local shops too: from juice to pastries, liqueur to cheese, hand cream to soap.

Thousand star hotel
To see Terschelling's real jewel in the crown you'll need to wait till it gets pitch dark. De Boschplaat, a 15-square-mile (40 km²) nature reserve on the eastern side of the island, has been officially named a Dark Sky Park by the international Dark Sky organization IDA. Terschelling is one of the 20

darkest places on Earth, in one of the most light-polluted countries in the world. Dark Sky Parks are usually located in natural areas where silence reigns and rare animal and plant species live. This is the place to experience real night-time darkness.

I set off on an evening stroll, or *joon walk* as they say in Aasters, a Terschelling dialect. We straddle dark and light. All you need for that is time. Because nightfall doesn't just mean danger, but also slowing down, connection, peace and space. All things we could use a bit more of in these hectic times. Simply sitting and watching the darkness rise, the boundaries blurring, the day fading. A small act of resistance against the absurd idea that every second needs to be spent usefully, and against the pressure of screen existence that controls us.

We shuffle carefully through the dunes. I hear a bat and a moth. The Milky Way is clearly visible. I don't need a telescope to admire the stars. Lines become softer, the night sounds clearer, the coolness clammier. It's wonderful what darkness does to your senses. Even the boundary between yourself and your surroundings dissolves eventually. I know that this piece of beach is quite small and must be close to the edge of the village, but surrounded by so much grey and black, I've lost my sense of direction. Everything seems to blur into one. A little imagination goes a long way in this shady world. //

Terschelling is one of the 20 darkest places on Earth, in one of the most light-polluted countries in the world.

Into the darkness.
Night-time images on the
Boschplaat.

I know that this piece of beach is quite small and must be close to the edge of the village, but surrounded by so much grey and black, I've lost my sense of direction.

The Drenkelingenhuisje seems to take on another dimension at night: a lookout post in a new universe.

LOCATION

Terschelling
The Netherlands, Europe

CURIOSITY

The Wadden Sea World Heritage Site is a unique tidal area: a world of ebb and flood. One moment everything is under water, the next moment you can walk on the seabed.

ADVENTURE TIME

The most spectacular time to visit is during autumn, when hundreds of thousands of migratory birds attack the tidal mud and sand flats.

TRAVEL TIP

Experience 'running aground' on a historic sailboat in the middle of the Wadden Sea. Sail the low and high tides, let the ship run aground and refloat it after a hike across the tidal flat. Don't forget your rubber boots and binoculars.

BEFORE YOU GO

Book a guided walk during the new moon, grab a star map and discover a whole other nightlife.

CHERNOBYL
51.2763° N, 30.2219° E

4.

An Exclusion Zone

Walking the streets of a nuclear ghost town

On 26 April 1986, the little-known Soviet city of Chernobyl – Chornobyl in Ukrainian – put itself well and truly on the map with what would turn out to be the biggest nuclear disaster in history. The power plant's Reactor Number Four exploded after a power outage test went wrong. The intense heat and pressure caused the roof of the reactor to erupt and a 1m (39 in) high yellowy green flame to shoot into the air. It would be days before nearby residents – and by extension the rest of the world – came to realize the disastrous consequences of this explosion and the radioactive cloud it caused. I'm going to Chernobyl, Pripyat, the purpose-built village of Slavutych and the region's abandoned sites where disaster struck about ten days after I was born. Lifting the lid on the cover-up.

Since the hit HBO miniseries Chernobyl, there has been a renewed interest in the disaster and region. The documentary The Babushkas of Chernobyl clocked up nine film awards. And the number of trips to the disaster zone increased by 40% between 2015 and 2020. But the way the abandoned Ukrainian sites were being run has created a negative undertone. As has the behaviour of certain influencers in this dark tourism hotspot. From inappropriate selfies and souvenir shops to 'exclusive rave parties', respect for the victims – officially, a few dozen; according to the UN, over 4,000 – seems to be in short supply. Was it time for a more introspective memorial? Probably we will never find out. In 2022 disaster struck a second

The rusty gondolas of the abandoned Ferris wheel in the exclusion zone.

time. Noboby ever thought this place would again make the headlines. Russian tanks, trenches in the Red Forest… : no dark tourist had ever anticipated that.

In the zone

We leave Kiyv at 7am for a 1.5-hour drive north. Long sleeves, long trousers and closed-toe shoes are the dress code. The Ukrainian summer might get hot, but you don't wander around Chernobyl in flip-flops. Eating outside is strictly prohibited. Briefly putting something on the ground is also a big no-no. Even if you're outside the most dangerous zone, the animals and the wind still carry around radioactive particles. So you're always on your guard.

These first important safety instructions were given to us by our guide, Alexander. He's only 22, but knows the entire exclusion zone like the back of his hand. He's been illegally urbexing – or urban exploring – since he was 16, and two years ago decided he may as well make it his job. His short sleeves and sense of humour reveal his determination. They'll never be rid of him.

We have to pass through several 'border controls'. Our passports are checked twice: once when we enter the 30-kilometre (19-mile) zone, and again into the 10-kilometre (6-mile) zone. The former is just a buffer zone: harmless, but necessary to avoid the transfer of contaminated particles or objects. In the zone, we are given a mandatory iris-sensitive badge to hang around our necks. Among other things, it measures our daily dose of radiation. Our eyes are apparently

47

our weakest spot. Without the protection of our skin, the radiation could work its way right through our bodies. The eyes of the firefighters who tried to put out the first fire in Reactor Four even turned blue. This dosimeter, which we diligently hand in after our visit, also allows the government to check whether visitors – and especially the guides – are respecting the set boundaries.

We also have our own dosimeter to measure the alpha, gamma and beta radiation we're being exposed to. Beta radiation is only found in the reactor itself. The numbers peak in the so-called hotspots: zones with dangerously high values. These are places where the wind or rain caused radioactive dust particles to gather, or where radioactive waste was dumped.

Ghost town walk

For the first hour, our group is almost literally tiptoeing around. When a butterfly flies into the van in the 10-kilometre (6-mile) zone, we try hard not to panic. Nobody wants to get anywhere near it, no matter how harmless it looks. But this isn't the only wildlife here. The excursions and trips to the zone not only attract so-called disaster tourists, but also wildlife spotters. Because in the evacuation zone, spanning 2,600 km² (1,000 square miles) on the Ukrainian side and 1,600 km² (620 square miles) on the Belarusian side, Europe's largest rewilding experiment is in full swing. These villages have been almost completely overgrown, turned into a jungle where wild wolves, horses, emus, lynx, bears and other endangered species roam free. There are 231 bird species to be spotted here.

The dosimeter ticks faster and faster and the alarm goes off for the first time. It's nail-biting, but nobody is actually tempted to bite their nails here. Along the way, Alexander keeps talking about 'the invisible enemy' that is still making life so difficult for the local people and rescue

workers. It's no use hiding or fighting back when you're up against this kind of opponent. It's an enemy that gets inside your head and under your skin, sometimes showing its bad side only decades later.

But this hasn't put everyone off. Today, 1,500 people still work in Chernobyl itself. The employees do shifts lasting from 10 minutes up to 10 days, depending on the zone they work in. They maintain, monitor and guard the nuclear reactor, which after more than 30 years still presents a hidden danger to humanity. Around a hundred unofficial inhabitants, or 'resettlers', have returned to the region. They're outlaws who live by their own rules, having decided to seek out their old homes and live completely outside normal society.

While chatting to us, Alexander mentions that he and his friends sometimes spend weeks camping in the zone. He and other 'stalkers' have furnished an entire flat in one of the abandoned apartment blocks. He won't tell us exactly where it is, but 'there's even a window in the frame'. A dog wanders through the abandoned sports stadium, which could easily seat 3,000 people. Alexander calls out to him: Alpha was very appropriately named after the most common radiation value. He follows shyly behind us as we continue walking. In 1986, abandoned pets were ruthlessly lured and shot because their radioactive fur was a danger to the people left behind. Today their offspring live in groups and in the wild.

It's hard to imagine that Pripyat was once a bustling city. We walk through and past abandoned theatres, a boxing club, restaurants and the legendary nursery. Not one

Street art by the Bane&Pest collective, who travelled to Chernobyl with 50 kg (110 pounds) of spray paint to create the first legal street art paintings.

In the evacuation zone, spanning 2,600 km² (1,000 square miles) on the Ukrainian side and 1,600 km² (620 square miles) on the Belarusian side, Europe's largest rewilding experiment is in full swing.

person got to go on a ride at the amusement park. It was due to open on 1 May 1986, just five days after the disaster. The big Ferris wheel now rusts in peace, but still looks like fun in the bright midday sun. Until Alexander warns us not to touch it. He points to a hotspot at the bottom of one of the pods. Our dosimeter peaks at values up to over 2,000. The helicopters that flew back and forth to the reactor are said to have emptied chemical products over this part of the park. Where they hit the ground or objects, the radioactivity peaks.

When the van turns off towards nuclear Reactor Four, we can hardly believe where we are. We can almost touch the door of the reactor. It's hard to imagine that, 36 years ago, the world almost ceased to exist here. Even rubbing your eyes in disbelief isn't allowed. On the way back, the guide shows us a gigantic abandoned Soviet over-the-horizon radar (OTH) system: a super-secret complex that would have cost more than twice as much as the nuclear power plant itself. The 150-metre (490-foot) high Duga-3 (code name: Object Chernobyl-2) was used from 1985 to 1989 to track intercontinental missiles with nuclear payloads. The range was so wide and the broadcast frequencies changed so often that it disrupted broadcasts worldwide. It was nicknamed the Russian Woodpecker because of the tapping noise it made on the short wave.

Emergency exit through the gift shop

Whenever we want to leave public cafeteria or zones, we, our clothes and our things are measured. Anyone with over 17 particles per square centimetre isn't allowed to leave the zone. The tyres and steps of the van are also scanned. At the 30-kilometre (19-mile) checkpoint there are two souvenir stands where a Chernobyl postcard, gas mask, white suit, sticker set or everyday dosimeter will cost you a tenner. 'Life is short, eat Chernobyl ice cream', reads the slogan above the ice-cream stand. The potential danger isn't hidden away, but rather embraced as their strongest selling point. The shops are so awkward, it's easy to forgive them.

The zone is an eye-opening place, which, in addition to its radioactivity, has an incredible beauty. The guides are also interested to see what impact the success of the HBO series will have on the place. It'll be another 300 years before the 10-kilometre (6-mile) zone is more or less safe for humans. And we'll only know the real impact on the people and the environment in tens, perhaps hundreds of years. Any official figures are either lacking or can barely be called reliable. My measured daily dose turns out to be no higher than that on a transatlantic flight. //

Wild fox in front of the Pripyat welcome sign. *(top)* **Abandoned school bus.** *(bottom)* **Control room of reactor four.** *(following pages)*

The zone is an eye-opening place, which, in addition to its radioactivity, also has an incredible beauty.

Soviet mural at the post office in Pripyat. *(right)* A bird's-eye view on the bewildered city center of Chernobyl. *(following pages)*

LOCATION

Chernobyl power plant
Ukraine, Europe

CURIOSITY

This zone represents the third-largest nature reserve in mainland Europe and has become an iconic experiment in rewilding. The land surrounding the plant, which has been largely off limits to humans for three decades, is a haven for wildlife.

TRAVEL TIP

Monitor Ukraine's current safety and entry requirements. When in doubt, consult your doctor for a medical check-up before departure.

BEFORE YOU GO

Watch HBO's five-part *Chernobyl* miniseries that tells the powerful and visceral story of the worst man-made accident in history, or read the graphic novel *The Lost Child of Chernobyl*.

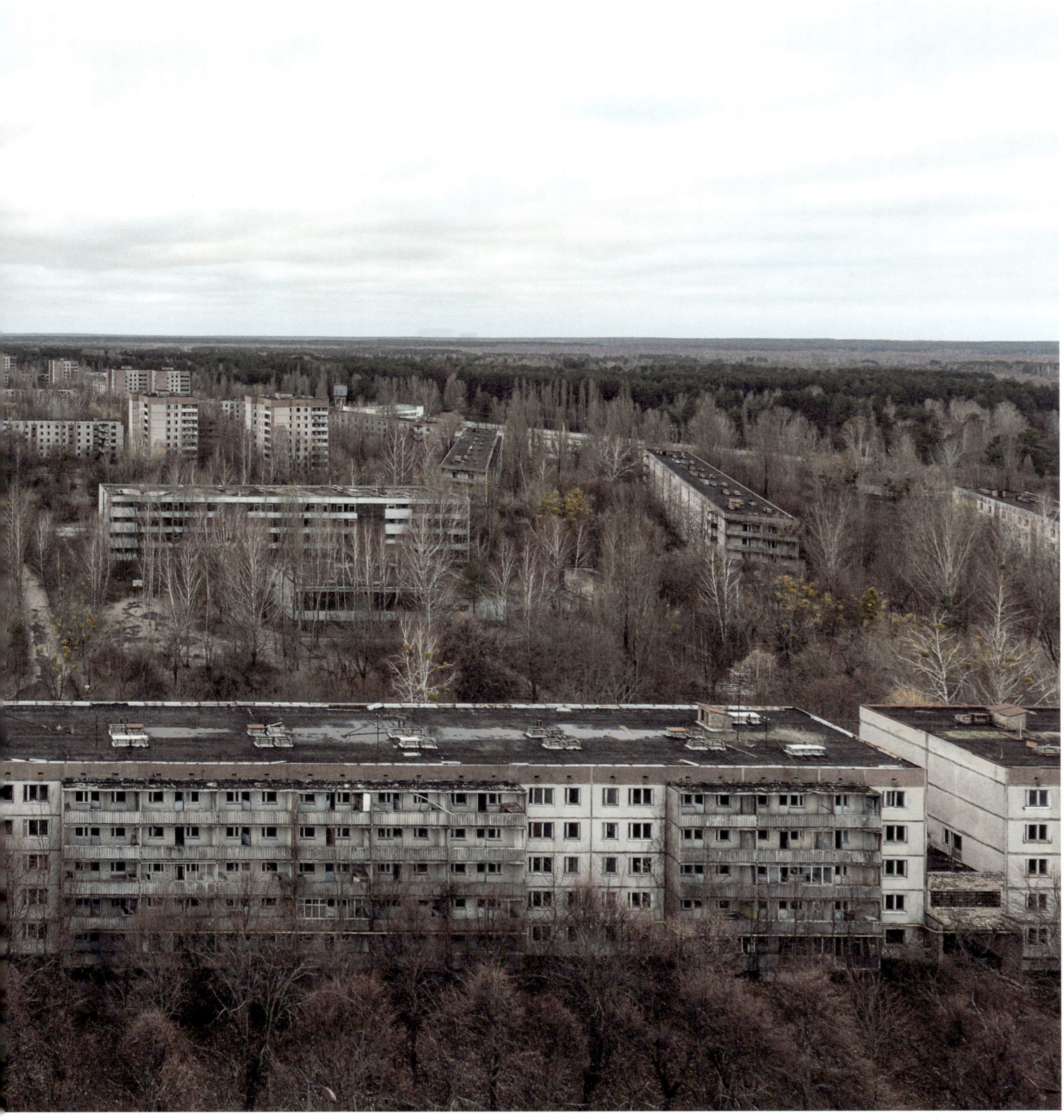

5

A Micronation

Transit through the country that doesn't exist

Transnistria is a country in limbo. Call it what you will – a micronation, twilight zone, robber state, Transdniestria, the Pridnestrovian Moldavian Republic or simply the PMR… There are plenty of names for the wannabe nation.

Officially, the country still belongs to Moldova, but in 1990 the Transnistrians declared independence in the strip between the Dniester and Ukraine. Following a turbulent period after the disintegration of the Soviet Union, peace has returned. There's football, singing and dancing. But the atmosphere is eerie in this 'country that doesn't exist'. It's not just the conflict but also time that seems to stand still here. A transit that stays with you.

It's the height of summer. We have a smooth border crossing into Moldova after our journey through Transylvania. There are no longer any official warning signs adorning these frayed edges of Europe. Almost all roads lead to the capital Chişinău, and the world's largest wine cellars. Although the city's history dates back more than six centuries, much of it was destroyed because of events in the Second World War and a major earthquake. The Soviets rebuilt the city and made it what it is today: a megacity of concrete skyscrapers. We're already getting used to Russian brutal modernism, the Cyrillic script and accompanying dialect. Just the rouble remains unchecked. A lot of the older generation have the feeling that things were better in the old Soviet times, while many younger people want to move on and build a future, often elsewhere. One in three Moldovans has left.

Tiraspol. Rehearsal for the grand parade to celebrate the 25th 'Independence Day' of Transnistria.

Our hotel is situated at the edge of the embassy district. The contrast between the Ladas in the city centre and the Porsche Cayennes that tear through the streets here couldn't be more pronounced. They're clearly not used to having a lot of foreign visitors at Hotel Nobel. We're greeted by a man with gold teeth, blue Crocs and a camouflage suit, and it looks like we're the only guests. Our room for three only has two beds. After a shower, the sink burps and the toilet bubbles over. Goldtooth makes up for it with a stack of pancakes and a tall glass of unspecified alcohol. 'An unmatched travel experience', says the hotel brochure.

Time travel

We drive a long way through Moldovan fields, towards the most heavily armed makeshift border: Transnistria. The country that doesn't exist has more than half a million inhabitants and is 4,163 km² (1,600 square miles) in size. The state is not officially recognized by the UN or Moldova, but it does have its own militarized checkpoint – complete with passport registration and entry permits. We need a special migrant permit, unless we stay a maximum of ten hours. We think better of asking what happens if we exceed the limit of that visitor visa. After waiting an hour and a half at the border crossing, we get a handful of stamps from an antique safe. When we're finally allowed to drive on, we cross the Dniester and zigzag past a few checkpoints, soldiers and tanks.

Since 1990, Russian 'peace troops' have been in charge here. The guys give us a friendly nod, but then stare at us a while. We try not to stand out too much, but keeping a low profile is difficult on our fully loaded touring motorbikes. The few tourists who visit the place rent a car with a driver so they can get around without being the ones to make a wrong turn. We're city-tripping in second gear. We don't stop at will, we hardly take any photos and we stick closely to the speed limits. In a self-proclaimed state, you'd rather not get on the wrong side of the police. Transnistria does have its own currency, anthem, flag and government. But without official recognition or an embassy, you're not entitled to foreign aid in this area. And as far as the insurance companies are concerned, you're an outlaw.

Football nation

In Tiraspol we ride between old Ladas, Lenin statues and busts of the president and Yuri Gagarin. Some sort of glitch in time has catapulted us to a 90s Soviet city. Trolleybuses are adorned with red stars, and buildings with large mosaics depicting Soviet scenes. In several former Eastern Bloc countries, symbols such as the hammer and sickle are banned. In Tiraspol you'll see them on every street corner, as in an open-air museum. This micronation is independent, but clearly relies heavily on Soviet sentiment.

The Transnistrian rouble also refers to this. The local currency is only available at local banks and currency exchange offices. The prices here are quite low, but there isn't much to buy. In the bookshop on the main road you can pick up PMR posters as well as kitschy souvenirs. At the post office you'll find stamps that are worthless outside the zone, but that reveal a lot about the identity Transnistria likes to assume. As do the Transnistrian flags that hang in every street next to those of the Russian Federation. A Russian tank has been put on a pedestal in the heart of the capital. 'For the fatherland' has been hastily scrawled on one side.

Other than this, there aren't any real sights. Sheriff is the largest (state) corporation in the micronation, and is ubiquitous. Petrol stations, supermarket chains, the TV channel, a construction company, advertising agency, mobile phone provider, car dealer and drinks brand all bear the name. As does the brand-new Sheriff Stadium, where FC Sheriff Tiraspol plays its home games. A neutral zone of 106 by 69 metres (348 by 226 feet).

Despite all the tensions between Moldova and Transnistria, one thing survived the declaration of independence: football. Through all the conflicts, the club from Tiraspol has played in the Moldovan league, where it is the record holder. During the Euro qualifiers, the national Moldovan team played against the Netherlands at the Sheriff Stadium. They suffered a narrow 2–1 defeat, but that was by no means the biggest intrigue of the evening. At kick-off, the Moldovan national anthem was played on Transnistrian soil, and both flags fluttered next to each other in a show of true sportsmanship.

Every year, teams from Chişinău also play some charity matches against Tiraspol. The proceeds go to veterans of the civil war conflict. Communities on both banks of the Dniester are always looking for ways to work together, even though the club's founder is a former Russian intelligence officer. 'There has never been any conflict between fans. Football is not politics. Football is for everyone. It doesn't matter if you're Moldovan, Russian or Ukrainian', pleads Pavel Cebanu, former president of the Moldovan Football Federation.

No-go zone

Eight hours later we cross the Ukrainian border. We are repeatedly urged to avoid Donetsk and Crimea. We spend the night in Odessa, at that point still a cosmopolitan free port on the Black Sea. Nothing to suggest that just ten months or so later there'll be warships anchored here in a city under siege. It's hard to imagine that residents will be forced to leave their pets behind in the zoo we walk past. After a morning dive, we decide to cross Ukraine via Kremenchuk towards Kharkiv and then into Russia. A detour of 90 kilometres (56 miles) that takes us more than five times further than expected. The inland road network is a tangle of asphalt with potholes and dried mud. On both the main and side roads, the asphalt bulges in all directions, like plasticine. This springtime mud season will later turn out to be the unexpected key to a military plan. Roads, days and encounters that we look back on today with a completely different perspective. //

The main administrative building in Tiraspol displays its communist emblems with pride.

Motor militia next to Russian flags during the rehearsal in the main street with sponsored pictures of the country in the background.

LOCATION

Transnistria
Moldova, Europe

CURIOSITY

In this breakaway state you can visit two pseudo-embassies – of the two other unrecognized countries South Ossetian and Abkhazia – to get a pseudo-stamp in a pseudo-passport.

ADVENTURE TIME

Due to reinforced military activity, including in areas close to some Moldovan borders, local travel advice should be monitored closely.

TRAVEL TIP

Finding the world's biggest Lenin bust can sound tempting, but I wouldn't recommend travelling into this state without a local guide. Visit the local bookshop when you have the chance.

BEFORE YOU GO

Before you travel, check Moldova's current safety and entry requirements.

ALTERNATIVE
WAYS

CANADA
50.1162° N, 122.9535° W

By Seaplane

Gliding over glaciers and lakes

6.

When you hear the words 'road trip', a strip of Canadian asphalt springs to mind. Teeming with highlights, the country's national parks are towards the top of any fly-drive wish list. It's very easy to explore the great outdoors here. In a camper van, along highways that lead you right through the Rockies past bears and glacial lakes. Yet Canada's true nature only really reveals itself with a bird's-eye view, away from the well-trodden paths inside the parks. And so halfway through we trade our motorbikes for a seaplane for a drive-fly that uncovers a different side to the country.

Our bikes will follow us, from Seoul to Vancouver, across the Pacific. This marks our twelfth border crossing since we left Brussels on our world tour. Since then we've already covered almost 20,000 kilometres (12,450 miles) eastwards overland. Getting a vehicle across a border always kicks off a flurry of paperwork, leaving me with a collection of stamps and clammy hands; if I lose them I might not get out of the country with my bike.

To get the bikes on board the plane, they first need to be packed to size. The transport costs aren't calculated according to weight, but by cubic metre. Cue some precision engineering to limit the dent in our budget. We disassemble parts and scratch our heads over mirrors, front forks and luggage.

A week later our bikes arrive all neatly packed, but getting them out of their boxes also turns out to be no easy feat. After managing to break open the box without any tools, we discover that our helmets and boots are mouldy. Yet we're happy to see them again: we

Seaplane flight over Garibaldi Provincial Park.

have another 10,000 kilometres (6,210 miles) of Canada and the US ahead of us. As we load up the bikes ready for departure, Vancouver softens the jet lag with its ocean breeze, views of snowy peaks and forest. The proximity of these natural elements lends a certain tranquillity to the port city. Even the city's skyline looks modest next to the surrounding mountain flanks. Being faced with a giant quickly puts things into perspective.

Animal park

We're immediately overwhelmed by the nature of Western Canada. On our bikes we cross the Rocky Mountains via the Icefields Parkway and the Columbia Icefield towards Whistler, stringing together the must-sees. 235 kilometres (146 miles) from Lake Louise in Banff to Jasper National Park are one of the world's most scenic drives. *National Geographic* agrees, categorizing it under 'Drives of a Lifetime'. Banff National Park itself is Canada's oldest, and was designated a UNESCO World Heritage Site in 1885. And this national treasure is clearly cherished. Despite the temperature fluctuations, there's not one crack or pothole in the road. Many of the broad and stately roads we're riding on were built for the 2010 Olympic Games.

The perfect highway that we're following clearly crosses a much older, natural highway. The 6,641-km² (4,126-square-mile) area is home to more than 280 species of birds and 50 species of mammals, including wolf, puma, lynx, eland and bear, which can roam freely and make use of the more than 40 wildlife crossings. The rivers are also teeming with life. In the autumn here you can witness the great salmon migration, where millions of salmon swim hundreds of miles upstream to return to their spawning grounds in the region's rivers and streams.

Canada's true nature only really reveals itself with a bird's-eye view.

But the area never feels truly wild to me. Even when we spot a grizzly bear by the side of the road, I barely flinch. Is it down to social media that Highway 93 feels a bit like Rue Déjà Vu? The kind of place where brochures and Instagram walls come to life? There's so much to see in such a limited space, but little to discover in a truly spontaneous way. We enjoy it, but something's not quite right. Around every bend there's another breathtaking landscape, with an info board and a designated photo stop plus a small crowd gathered. For the first time in months we feel more like tourists than travellers. With its four million visitors a year, Banff is almost becoming a victim of its own success.

Continental divide

Fortunately, our wheels keep us focused. Although we're not always in our comfort zone. We've been living out of our panniers for 70 days and wearing the same handful of clothes all that time. On the road, the icy wind keeps us alert, and each of the five climatic zones we cross through introduces itself with its own soundtrack and scent. In the evening we pitch our tent next to the horse-watering hole of a campsite that has been fully booked for months. We continue with our stubborn approach to travel – no making of plans or reservations – but this is the first time we need a pass to be allowed to see the great outdoors.

The appeal of this region isn't new. It's been an attraction for a century and a half. For nature lovers, but also for gold and fortune seekers of the Crown Colonies. Ten years after British Columbia joined the Canadian Confederation, construction of the transcontinental railroad through the Rocky Mountains began in 1881. The Canadian

Pacific Railway (CPR) opened up the Banff area to a large audience. It became a gateway through the mountain range that previously split the continent in two over a distance of nearly 5,000 kilometres (3,110 miles).

The area first became popular when some railway workers hit upon sulphurous hot springs at the base of Sulphur Mountain. The Banff Springs Hotel opened its doors there in 1888. It was the place to be seen among the Victorian nobility, who loved to come and feel rejuvenated by the 'healing waters'. We freshen up back at the water pump.

Expo 86

In an attempt to get an alternative view of the region, we ride the Sea to Sky highway to the seaplane base on Green Lake in Whistler, hoping for a maiden flight in the country that likes to show off its special aircraft. Because here and there you can still see remnants of the World Exposition on

Unusual catch of the day. *(left)* **Rider's high on the Columbia Icefield.** *(right)*

Here and there you can still see remnants of the World Exposition on Transportation and Communication, or 'Expo 86' for short, which Vancouver hosted in 1986.

Transportation and Communication, or 'Expo 86' for short, which Vancouver hosted in 1986. It was the first world exhibition on a transport theme. The three interlocking rings that form the logo's number 86 represented the three main modes of transport: land, sea and air. The expo coincided with the city turning 100, and the 100th anniversary of being able to reach the west coast via the Trans-Canadian railway. The world was introduced to Vancouver through monorails, ferries, gondolas, a Japanese 'SkyTrain', kites, hot air balloons and double-deckers.

Yet it is the bad timing of the US and USSR pavilions that many visitors remember the most. The US was promoting the American space programme just a few months after NASA's Challenger exploded. Meanwhile the USSR was boasting about their thriving nuclear industry, barely a week after the Chernobyl disaster.

Workhouse of the north

Despite all the futuristic attempts, it is the historic seaplanes that steal the show in the Canadian outback today. At Harbour Air, one of the largest all-seaplane airlines in the world, we get a 'crash' course in

Seat with a view of the Blackcomb Glacier. *(left)* Right through the Rocky Mountains. *(top right)* Improvised camping at the horse-watering hole in Banff. *(bottom)*

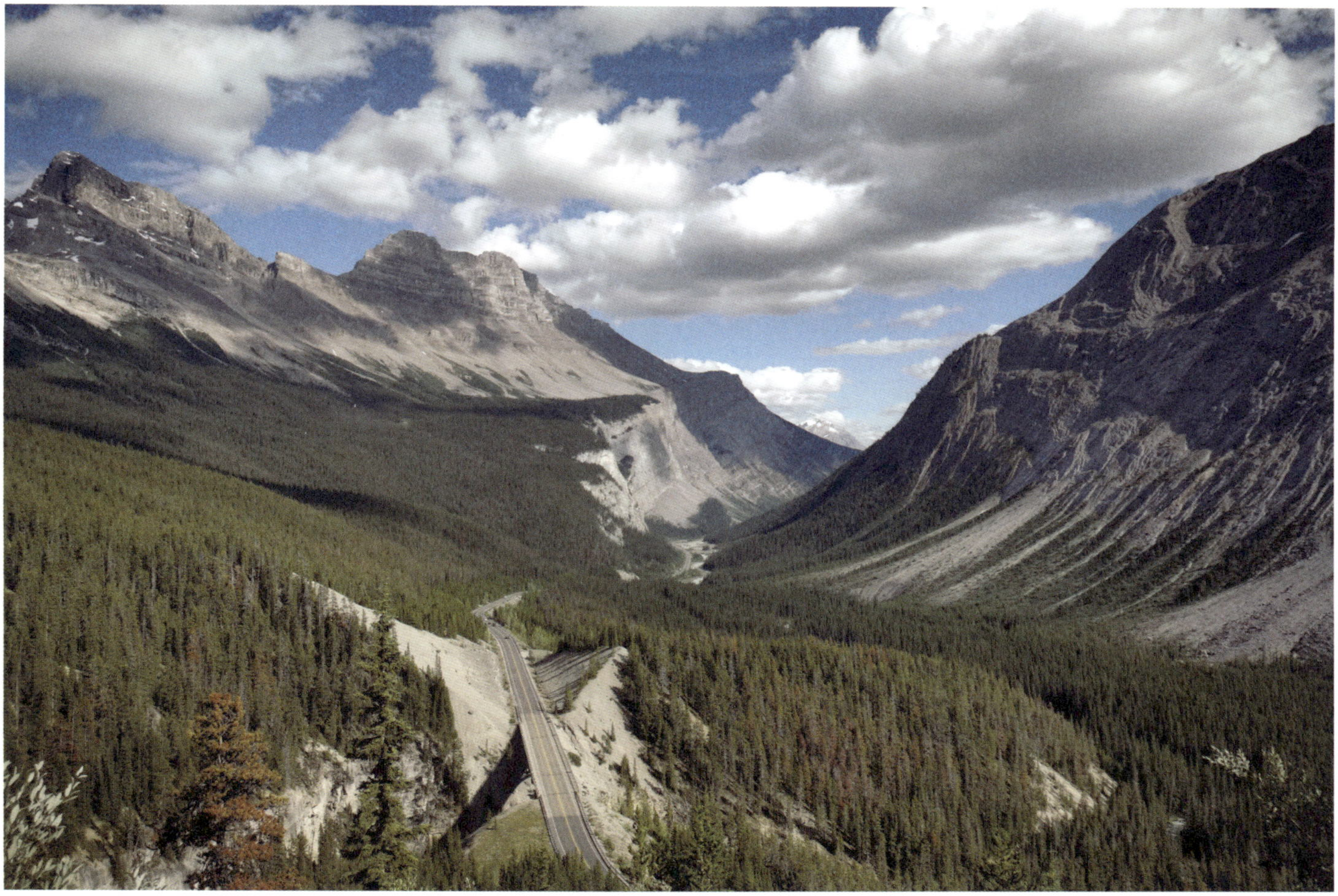

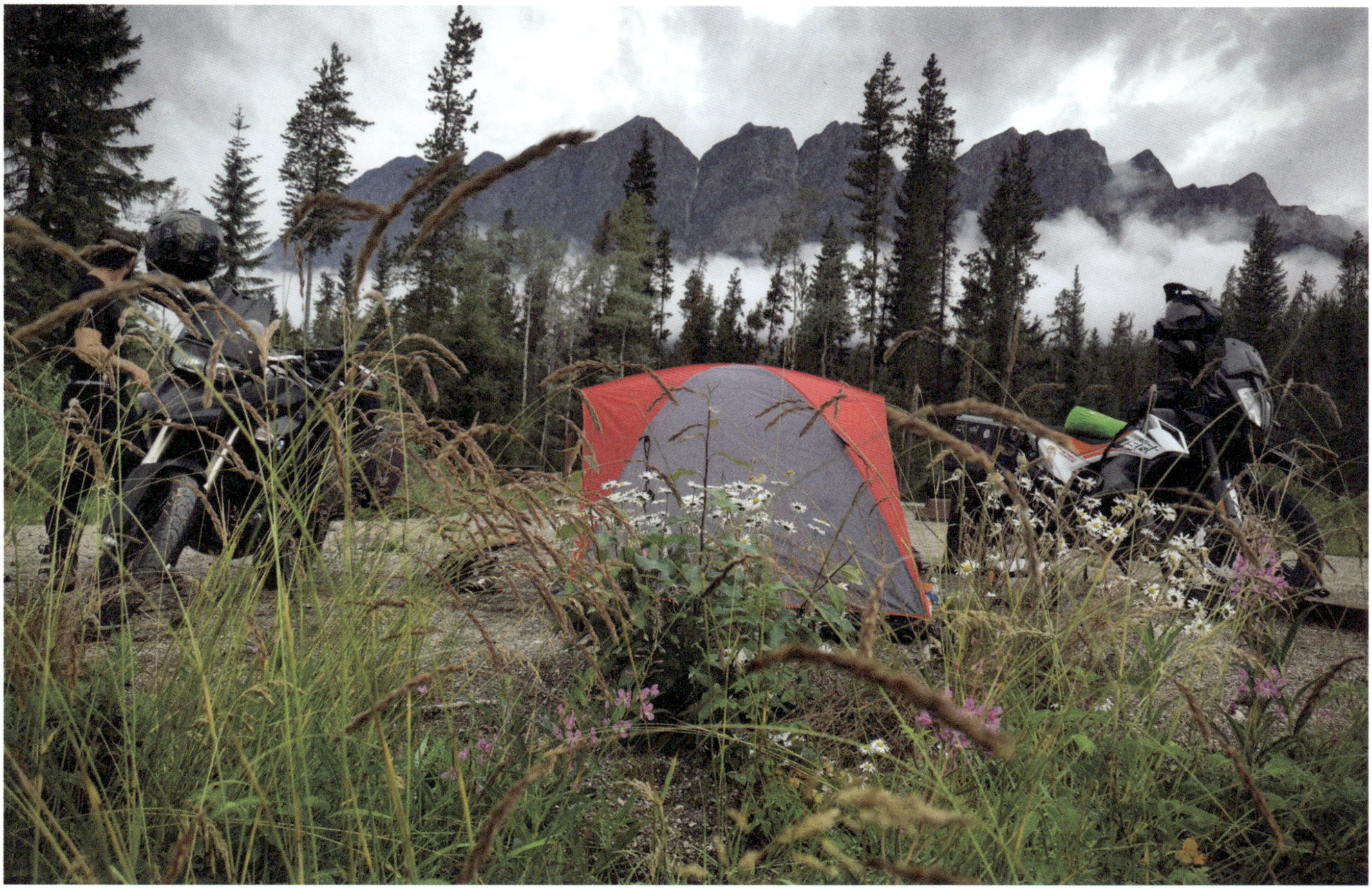

floatplane flying. Seaplanes were especially useful before the Second World War, when there weren't that many airfields. The first seaplane trips took place in the most unpredictable conditions. With no knowledge of the terrain, weather reports, radio assistance or recovery or landing sites, the pilots traversed and photographed miles of unexplored wilderness in a matter of hours. Something that would take weeks by canoe, horse, dog sled or on foot. So the seaplane played such an important role in the development and exploration of wilder Canada.

Seaplanes are often the only lifeline between off-grid communities and civilization. Flights run a few times a week from several villages on Vancouver Island, bringing parcels, people, animals and 'anything that fits' to the most remote areas.

As well as floats, skis – instead of wheels – are also used in winter. Sometimes it gets so cold in certain places that they even take the oil out of the engines, store it inside and pour the warmed-up fuel back in before they set off again.   **Sunset over Lake Louise.**

Sometimes it gets so cold in certain places that they even take the oil out of the engines, store it inside and pour the warmed-up fuel back in before they set off again.

Reflections and roaring engines at the seaplane base in Whistler.

The engines are now roaring and the pilot gives instructions through the headphones. We taxi over the water and just a few seconds later we're hovering over Garibaldi Provincial Park. A spectacular glacier tour, overlooking a nature reserve full of volcanic peaks, snow-capped mountains, turquoise lakes and mountain meadows.

It is not just beauty and splendour that the Blackcomb and Horstman Glaciers reveal. Over the past 125 years, these glaciers have become half as thick. Whistler is one of Canada's most popular ski destinations and the largest ski resort in North America. But maybe not for much longer. Climate change, increasingly shorter summer ski periods and retreating glaciers are putting pressure on the local ecosystem and communities. And Whistler is taking that responsibility seriously. Through various eco-campaigns and a 'Commitment to Zero' policy, the region is taking on a leading role in 'regenerative tourism', where the idea is to leave a place better than you found it. The local communities are once again prioritizing nature, experimenting with new energy systems, hybrid snow trucks and recently unveiling a world first in aviation: the first fully electric seaplane fleet. A new era of nature-first adventures? //

LOCATION

British Columbia
Canada, North America

CURIOSITY

Look up in Vancouver, and you'll see something unique about the skies overhead: seaplanes are everywhere. A lot of Canadians reside in hard-to-reach places, inside canyons and beside towering mountains and glaciers. Building a landing strip in those areas isn't possible or economically worthwhile. But a seaplane can land pretty much anywhere that's wet.

ADVENTURE TIME

Spring (March to May) and autumn (September to November) are the shoulder months, when there are fewer tourists and the weather is pleasant. Whistler Floatplane service connects Victoria and Vancouver to Whistler from May to October.

TRAVEL TIP

Book a glacier trip with Whistler's Harbour Air, and fly over majestic glaciers, peaks, volcanic formations and lush alpine meadows.

BEFORE YOU GO

Unfortunately wild camping isn't allowed in the national parks, and campsites are often fully booked way in advance. But if you don't plan ahead, you're in for an adventure.

On a Surfboard

Catching waves in the demilitarized zone

7.

Neon lights, pilgrims, watchtowers, robots and Buddha's hand: South Korea is a peninsula of extremes. It is artificially cut off from the continent to the north, and splintered into some 3,000 islets in the southern Yellow Sea and the Sea of Japan. This interim paradise is officially still at war: even after 70 years of ceasefire, the border with North Korea remains heavily fortified.

Artists and surfers come in search of freedom here, between the military fortifications on the east coast. Overlooked by cameras and military outposts, they find their muse as well as North Korean cigarette butts in the waves.

I say farewell to Russia at a dock in Vladivostok. Our two motorbikes have been loaded onto the ship and the foghorn of the *Eastern Dream* heralds the end of the trans-Siberian leg of our journey. We have already covered more than 20,000 kilometres (12,430 miles). On land, Lenin shows us the way east one last time. Although the ship is still scraping against the Russian shore, the footbridge already feels like a border crossing. On board, upbeat Asian tunes fill the air and Japanese and Korean products are offered to us with a nod. The clash with masculine Russia couldn't be any bigger. Everything here is already *kawaii* – the Japanese word for 'cute'. We each choose a free capsule bed in a cabin for 60 people, where Russian, Japanese and Korean chatter blend into one. The passengers are a mixed bag as colourful as the interior.

There's a problem with one of our bikes. We'll have to see how and where we can get it up and running again. That surrender to the unknown creates a new kind of peace of mind

on long journeys. And rightly so, because soon enough the solution presents itself to us in the guise of a Korean biker. Hong Jeong Soo had spotted us at the migration office earlier that day and wants to talk about motorbikes. When he hears about our technical issue, he immediately offers us a solution for when we dock. His friend Yeon is a mechanic in the harbour. We make a pact and it's like shaking hands with a sumo wrestler.

Explosion of colour

Not even 24 hours later, we dock in South Korea. We leave the broken bike in Donghae, where the generator will be rewound by hand. Jeong leads the way as we follow on one bike. Before long, he stops in Gangneung, one of the host cities of the 2018 Winter Olympics. Here, North and South Korea provided the biggest sporting twist in decades when they took to the stage under one flag at the opening ceremony, and even formed a joint ice hockey team. However, the rapprochement didn't go much further than a bit of 'sunshine symbolism', explains Jeong as he conjures up a banchan table for us. A loaded welcome, with kimchi, rice, bulgogi BBQ beef, seaweed, soup and tofu. It's been weeks since we've eaten food this colourful.

Gangnam style

Inland, the scent of Little Trees air freshener and cold brew coffee wafts towards us. South Korea has a lot of similarities with Japan, but seems more relaxed and less disciplined. It's like a polished miniature version of the real world, with friendly shapes and colours that make even police checkpoints and military bases

Surfer on guard at Yangyang Surfyy Beach.

look cute. This takes a bit of getting used to after a rougher trip through Mongolia and the former Soviet Union.

Riding into Seoul takes hours. Compared to the empty landscapes of recent months, the capital feels colossal. The roads are a maze of one-way streets, with old, new and retro-futuristic dimensions. Google Maps is banned here, and none of us reads Korean. It takes us days to make a first blueprint of this sweltering city with its countless layers, museums, hip neighbourhoods, pop-ups and temples. Yet, despite its 9.7 million inhabitants, Seoul never really feels overwhelming. Even the hustle and bustle seems to be waning in the heat.

Poetic freedom

Barely an hour's ride from Seoul, we enter another concrete beehive. Heyri Art Valley is an artists' village and community where more than 400 painters, photographers, musicians, architects and writers live and work together. The sight of barbed wire along the main road to Paju is the only clue that North Korea and the world's most controversial border zone beckon just 6 kilometres (3.7 miles) away. The demilitarized zone is a buffer between two armies, a piece of no man's land 4 kilometres (2.4 miles) wide and no less than 240 kilometres (150 miles) long. Since 1953, peace has never been officially declared, so North Korea still regularly provokes South along or over this line.

Writer and publisher Kim Eun-ho deliberately chose this location, along the 'Freedom Highway' in the shadow of the totalitarian regime on the other side, for his artists' village, which he completed in 2001. Here, he built a liberal open-air studio that is all about continuous creation. A place with no rules, walls or fences to hide yourself or your work behind. Eun-ho named the

The photo booth: one of Seoul's most photogenic souvenirs, both sides of the curtain.

village after a traditional harvest song, 'Heyri Sound', which farmers from Paju sing while they work in the fields. At the exact time more and more North Koreans were attempting to flee the country, in a cry for freedom in the face of famine.

We stroll through dozens of little galleries, mini museums, cafés and restaurants, all of which look unique. On the ground floor you get an insight into the creative process; upstairs is where the artist lives. Each human construction is built with respect for everyone's individuality and for nature. For example, constructions made of ecological materials are built around the natural elements, instead of the other way around. In the evening the street lamps are dimmed so you can still see the stars. New streets follow the contours of the hill. Ingeniously simple.

Peace and waves

Along the east coast, new communities are also popping up in search of a new kind of interaction. Not far below 38th Parallel Beach, the official demarcation line between North and South Korea since the end of the Korean War in 1950, night patrols switch places with morning surfers donning wetsuits and paddling into the ocean. We swap the bike for a longboard, pick up a handful of tips using sign language and dive into the Sea of Japan together with a few locals and Chinese tourists. The waves are too small to make any speed, but it certainly doesn't ruin the fun. The surf scene may be new, but the enthusiasm is ripe.

The east coast of Korea might not be the first place we westerners think of when it comes to surfing, but that may soon change. This surf spot was only recently 'discovered' because before that a lot of the beaches with good conditions were part of the inaccessible military zones. Since 2015, they have been cautiously opened up, albeit still closely guarded and with a strict ban after 10pm. It's still an absurd sight,

Surf tips from local Seung. *(left)*
Capsule beds on the Eastern
Dream. *(right)*

Not far below 38th Parallel Beach,
the official demarcation line between
North and South Korea since the end
of the Korean War in 1950, night
patrols switch places with morning
surfers donning wetsuits and
paddling into the ocean.

watching the surfers stepping over the barbed wire. The
artillery base is just a short distance away. Sometimes North
Korean cans or cigarette butts wash up after a storm, like items
in a time capsule from the world on the other side.

With sun cream smeared on their cheeks like war
paint, the surfers might seem ready to fight, but they're
longing for a peace treaty. 'Before, I never felt the North could
be a friend', says local surfer Kwon Min-ju. 'Now I feel very
close to it. In the end, all we want is peace and waves.' //

Drifting within sight of the
night patrols on 38th Parallel
Beach. *(previous pages)*
The Book House Foresta and
other curiosities in the
artists' village Heyri. *(top)*

Korean War Memorial. *(top)*
Catch of the day. *(bottom)*
Getting lost in the streets
and Leeum Samsung
Museum of Art in Seoul.
(following pages)

LOCATION

Demilitarized zone
South Korea, East Asia

CURIOSITY

The 38th Parallel Beach is a harbour, a military base, and a fast-growing hub for Seoul's young surfer class. The east coast of Korea has some of the most consistent and powerful waves in the country. Now that beaches around the demilitarized zones are opening up for the public, these waves are no longer left unsurfed.

ADVENTURE TIME

Despite the cold weather, the best time to surf in the country is during the winter months. From November to March the wind blows offshore, so you will get the best quality waves.

TRAVEL TIP

Book an artistic stay at the Heyri Art Village in Paju, best explored in 1-2 days.

BEFORE YOU GO

Watch the documentary The Winter Surf and discover the surf communities in Korea.

Evening port in Donghae.

In a Packraft

Navigating whitewater rivers in a portable boat

8

The French Hautes-Alpes is a popular destination for outdoor enthusiasts, especially water sports lovers. Whether you're just dipping your toe in or you're a seasoned adventurer looking for the latest thrill... the Alpine waters here have a challenge in store for everyone. A playground for rafters, kayakers and riverboarders.

Even before we unpack our tent, we can hear what we've come for. We set up camp on the right bank of the Durance, which today isn't bright blue, but a murky brown. The twinkle in the eye of our guide Servaes Timmerman reveals that this is actually good news, as it means there was a lot of fresh rainfall last night. The Durance rises in the Alps and flows more than 300 kilometres (180 miles) further into the Rhône, near Avignon. From early spring until autumn, it is filled with an abundance of meltwater from the surrounding mountains, and from the tributaries of the Écrins and Queyras National Parks. This makes it one of France's best-known rivers in terms of volume. Hello there, gushing white water! We're ready for our first adrenaline rush.

From north Alaska to the southern Alps

As we're unpacking we get a good look at each other's gear. As well as a tent, sleeping bag and other camping equipment, everyone also has a packraft: a compact single-person raft that rolls up. That's 2 kg (over 4 pounds) of inflatable freedom, in almost every colour of the rainbow. The outdoor kit to help you overcome the natural boundaries of the world. After all, when you're out in the wild you're often travelling on either land or water. For many backpackers, lakes and rivers stop them in their tracks; for rafters and kayakers, it's the land. Enter the packraft.

This innovative raft has its roots in wild northern Alaska, where there are more rivers than roads. Anyone who sets off on foot without any special gear is soon brought to an icy-cold standstill. Early adventurers got around this by using the heavy-duty Air Force rescue boats you could pick up after the Second World War. Or dinghies, which on a trip lasting several days would constantly need patching up. When Sheri Tingey's son kept returning home with one of these in shreds, she decided it was time to come up with something better herself. The first commercial Alpacka raft was born.

The idea of carrying a small boat with you to cross rivers and lakes certainly isn't new. Around 900 BCE, for example, troops were already venturing past their enemy on greased, inflated animal skins. The most stylish design resurfaced in Victorian England in the form of the 'boat cloak' by Royal Navy Lieutenant Peter Halkett. An ingenious coat-boat combo that he used to explore the Thames and the Canadian Arctic. His coat quickly turned into a boat, his umbrella into a sail, and his walking stick into a paddle. Row, row, row your coat...

Sink or swim

Whether we look quite as stylish, I'll leave to your imagination. I inflate my packraft using the inflation bag, and squeeze myself into the wetsuit. The air temperature is tropical, but the water has come straight from the glaciers. It's not just nerves making us tremble here. We apply factor 50 to our hands and noses. Then I put on my helmet, life

Guil descent along the Château Queyras with its powerful water suction.

jacket and neoprene boots extra tightly. With boulders and sharp rocks turning the river into an obstacle course, it's best to be prepared. A throw rope to rescue others from nail-biting situations in the more difficult sections is certainly not redundant.

We exchange a nod. It's showtime. Gulp. Just one more pee perhaps? I give my packraft one last top-up from the depths of my lungs, then push myself off the bank. Once on the water, the packraft wraps around my waist like an extension of me. The spray deck connects my torso to the raft via an elastic cord. I fly into my first bend and dodge an obstacle with a firm hip movement. As I crash into a wave nose-first, the raft keeps the water out, but I'm now on high alert. Through the thin skin of the packraft I feel every wave, boulder and branch. The raft is a lifebuoy, the paddle a direction indicator, which we use to forcefully carve our way through the river.

The water is unpredictable, varying from shallow to as much as 16 metres (52 feet) deep. The river carries us on its back, and with every surge we learn to read the water better. What looks safe can be treacherous, and vice versa. One minute your bum is scraping the rocky riverbed; just seconds later an unstoppable force catapults you down. One of us gets tangled up in a clump of low-hanging branches.

Someone else gets a paddle in the face. Another capsizes and tries in vain to swim to shore. I manage to dodge a boulder, but only just. It's an adrenaline rush, but it also crystallizes our focus. We'll still be grinning days after we've dried out.

A wild ride for everyone

We get throw-rope training, practise our knots and paddle the Durance, Ubaye, Rabioux, Argentière, the Sunshine Route and the Guil, including its Triple Chute and legendary Château Queyras gorges. In the evening our wetsuits dry in the sun and we talk for hours about the water, beer in hand. We dream about 15-metre (50-foot) high free-flowing waterfalls. The days fly by as we rapidly learn and get more and more confident in the water. Overconfident, even, as I also end up doing a lot of swimming. Fortunately, our experienced guides understand the water and weather, and are good at assessing the dangers. The situation can change from white-water class W3 to W5 or vice versa in no time, and then you really need to know what you're doing.

Like white-water kayakers, packrafters are extremely respectful of the mountains and rivers. Great partnerships then form between the community and local nature conservation organizations. We're all water. Whether we're crushing white water or enjoying a lazy float. //

Warm-up fun and material
check. *(top)* Ride home with
the white-water crew.
(bottom)

LOCATION

The Alps
France, Europe

CURIOSITY

The valley of Château-Queyras is a mythical spot for wild water
rafters. The Guil rapids are a must in the Alps: they cross the
Queyras valley and have forged a reputation for being one of
the hotspots for white-water sports in Europe.

ADVENTURE TIME

The water level largely depends on the season. The best time for
white-water rafting is between April and September, which is
when most of the companies offer their rental and tour services.

TRAVEL TIP

In July the French Alps Packrafting Meet-up takes place on the
Durance River, coinciding in part with the Durance
Whitewater Festival, which unites all river lovers, with kayaks,
rafts, SUPs and packrafts.

BEFORE YOU GO

Rent a packraft and have some paddle practice before
departure. *The Packraft Handbook: An Instructional Guide for
the Curious* is a great guide for beginners.

'By perceiving ourselves as part of
the river, we take responsibility for
the river as a whole.'

Václav Havel

Navigating between rock
formations in the Queyras
National Park. *(right and
following pages)*

In a Zodiac

Expedition training in open water

9

Like many adventures, this story starts with a nice coincidence and a simple 'yes'. Before I know it, I'm a ship's apprentice in Scheveningen harbour. Taken under the wing of an expedition leader and maritime photographer, I'm ready for some intense expedition training, being bashed about in an Rigid Inflatable Boat (RIB) and Zodiac by the swell of the North Sea. The fresh salty air, the wind in my hair. Forget everything on dry land for a while.

Katinka Puglia welcomes us with a coffee and a view of the harbour. But it's not just our instructor for the next few days who has a memorable name. Her list of former students also has international allure. From astronauts to helicopter pilots, water police and rescue workers: half the world has stepped aboard here. Katinka herself has navigated between elephant seals and penguins in remote places such as South Georgia. This week, our goal is to obtain our Royal Yachting Association Powerboat Level 2 certificate (International Certificate of Competence), an international licence required on expedition vessels, which mainly sail in Antarctica and the Arctic.

Katinka and her husband started more than 25 years ago as voluntary lifeguards at The Hague's Rescue Brigade. They learnt to understand and respect the power of the sea early on. On the water they were often able to provide assistance and save several lives. They wanted to share that knowledge. Their RIB School is therefore not just a sailing school, but one of the two international sailing schools in Europe where training is given on Zodiacs and RIBs. These types of inflatable boat are mainly known as lifeboats because they are seaworthy, fast and easy to manoeuvre, and can even land on the beach. The coastguard, navies, Greenpeace and Sea Shepherd – and, oh yes, James Bond – are known to use these RIBs. If you ever see a yellow coastguard rescue helicopter chasing a yellow RIB off the coast of Scheveningen, it's probably Katinka on board. From her boat she teaches the gentlemen a thing or two about hand-eye coordination. With her eyes closed.

First manoeuvres with a Zodiac in the harbour.

Get your bearings

Today we get our baptism of fire. Luckily I'm not completely unprepared. I got my General Day Skipper Steering Certificate in 2013 when a friend and I converted an old Vaurien (a small sailing boat) into a motorboat. My land-dwelling brain is running at full speed. We get a crash/refresher course on the moon phases and the tides, and learn to read the compass and the water. It's a lot to process in a short time, but from our smile you can tell we're having fun. This is the language of the first explorers and adventurers, who – when they needed to – were even able to navigate purely using the stars. A skill that still comes in handy today. Did you know the magnetic north pole is shifting every year? The magnetic north pole, which scientists believe should be over Canada, really is on the move. It's heading for Siberia at lightning speed – with disorienting consequences for the users of both old-fashioned compasses and new navigation systems.

Plan overboard

Time to put the theory into practice. We wrap up well. It's the height of summer, but it can be deceptively cold at sea. While the people next to us lounge about in shorts, I put on my expedition suit and pull my woolly hat over my ears. They

In our mind the water is 10 °C colder and we're not navigating between buoys and beacons, but between ice floes and whale tails.

don't bat an eyelid, so deep is the bond between the sea and many Scheveningers. In the morning here you need to dress for two to four seasons in one day.

The instructor immediately shows us everything in the Zodiac. From how to work the pressure valves to the layout of the air compartments... and her ingenious Shewee, which she made herself from an old ketchup bottle. I can already feel it: we're going to get along just fine.

On the first day we learn to sail the RIB using the steering wheel and throttle; on the second day the Zodiac using the tiller steering. In the Zodiac, you stand with your feet apart and keep your balance by putting enough tension on the rope on the left and the tiller on the right. I notice straight away that this requires my full concentration, because the throttle works completely the other way around on a motorbike. Balance, then gently does it. And so it's a slow start. There's a good reason they call the propeller the 'meat cleaver'. Slow is pro.

When I have the boat sufficiently under control, I can finally open the throttle. As soon as the boat picks up enough speed, it lifts itself up onto the three ribs. On a Zodiac or RIB, the bottom is made of a material that keeps its shape – often aluminium or polyester. A thick, air-filled rim, called the 'tube', surrounds the hull. A RIB cuts through the waves, while a dinghy goes over them. It's like we're flying.

Smooth sailing

The conditions are perfect: green water, white horses, a salty south-westerly breeze and a bright blue sky. The RIB and Zodiac that keep us busy for hours were christened *Humboldt* and *Shackleton*, respectively, after the world's most famous polar explorers. Ernest Shackleton's famous survival story, in particular, is a constant source of inspiration. Of course, the water behaves completely differently here than in Antarctica, where the strong wind calls the shots. That's why we seek out the wind as much as possible, from windward to leeward side. Mooring, flanking, beach landings, fast cornering and man overboard exercises. To the rescue! Practise, practise, practise.

During the training, Katinka keeps up the pace. We throw ourselves into the exercises, which are much more tiring than expected. They demand all your concentration and you often have to do four new things at the same time. In our mind the water is 10 °C colder and we're not navigating between buoys and beacons, but between ice floes and whale tails.

The Flying Dutchman

There it is at last: the turbulent North Sea. When Katinka thinks we're ready for it, we sail out of the harbour and immediately feel the power of the waves swirling under the boat. The wind cuts like a knife and I get a sudden splash

First Mate Mike clears the way for the three-masted *Stad Amsterdam. (left)*

Fishing boat in Scheveningen. *(top)* **Needs must: a Shewee self-made from an old ketchup bottle.** *(bottom)*

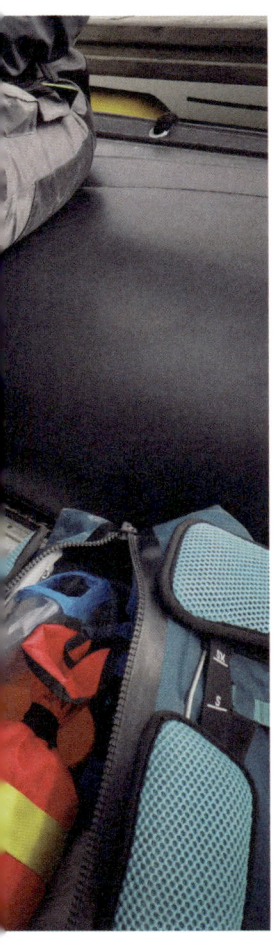

of seawater in my face. Fantastic! Read the water and learn to cut right through the waves rather than go around them.

After the intensive exercises and tests, we pass the training. We celebrate appropriately with a Scheveningen Bomschuit IPA from the tap and fresh kibbeling fried fish nuggets. In the evening I wobble a bit in my chair. Ready for the next chapter. And that would come much sooner than we could have hoped. Another step closer to that wild childhood dream. The adventure comes as it presents itself. Seemingly random, in fits and starts. Just like life itself. //

LOCATION

North Sea
The Netherlands, Europe

CURIOSITY

This training course is a first step to prepare yourself to become part of an expedition team of vessels sailing to either Antarctica or the Arctic.

ADVENTURE TIME

These boats perform well in all weather and sea conditions, if you know how to handle them. If the sea permits, you'll ride out to the cardinals at sea, and take compass bearings.

TRAVEL TIP

Take a stroll on the famous pier in Scheveningen after your training. With its Ferris wheel and cosy beach bars, it's a landmark on the Dutch coast.

BEFORE YOU GO

Watch the epic documentary *Shackleton's Endurance: The Lost Ice Ship Found* before you jump on the Zodiac carrying his name.

Man overboard manoeuvre in the harbour.

On a Vintage Motorcycle

Road trip on the roof of the world

I was keenly aware that I had to be careful. Adventure without risks is Disneyland, I had jokingly reiterated. Gearing up for the highest drivable pass in the world, with temperature fluctuations of 50 °C and 2,750 kilometres (1,710 miles) of open road. A motorbike journey from Delhi to Ladakh and not quite back again.

Nowadays, what stands in the way of a dream becoming reality is usually just technicalities. Not even six months after articulating the idea of riding to the eternal snow on an old-timer, the train ride from Delhi to Amritsar gives me a day to acclimatize. Outside, the heatwave scalds and melts the tarmac; inside, the air conditioning is on full blast. Here the price of a ticket is in inverse proportion to the temperature in your carriage.

Looking out of the small train window, India already reveals itself as a patchwork of everything. Landing in Delhi was just a warm-up for the onslaught of colours and flavours to come. I can't wait to hear the roar of the bikes through the mighty mountains. Tomorrow, I'll pick up the motorbike and join a small group of travellers. But for now I move in sync to the rhythm of the tracks.

Foothills in Tibet

I only need a few miles to find out what the expression 'sweating bullets' means. The Royal Enfield Bullet catapults me without warning into the merciless city traffic of Amritsar. The roads are bursting at the seams with trucks, buses,

The desolate road to Kargil, right through the valley of Zanskar.

wandering cows and the occasional monkey. Road markings appear distorted, as they have been melting due to the heat. We are supposed to drive on the left, but everyone actually does whatever they please. No, our most important safety measures won't be our brakes or helmets. Here, it is your horn or your life.

Luckily the refuge of the rice paddies beckons. In the small villages on the way to Wagah, children call out to us, waving wildly while forming peace signs. However, this area has been far from peaceful through recent history. This town is the only land border between India and Pakistan on the Grand Trunk Road. It has been separated by Asia's Berlin Wall since Pakistan's independence in 1947.

We continue our journey on the winding road to McLeod Ganj, on the foothills of the Himalayas. The 1,800-metre (5,900 foot) climb is not only noticeable from the heat on our brakes. From now on it becomes much colder in the evenings, crisp bags explode and two sips of beer are enough to get you quite tipsy.

We sleep in McLeod Ganj, also known as Little Lhasa. It's in this suburb of Dharamshala that the Tibetan government is based, and in it the Dalai Lama has lived in exile since he fled China in 1959. In the garden of the cloister, trainee monks learn philosophy and study the Tibetan Five Wisdoms during lively debates. They underline their arguments with big steps and powerful gestures, while they fire hypotheses at each other. Envious of so much devotion, I turn the prayer wheels to say a mantra, but I'm almost immediately

In a country where the air vibrates from the heat, snow is the most beautiful status symbol.

reprimanded. Here you only contemplate clockwise. When I leave the Debate Courtyard, it is in particular the echo of the empathetic clapping that stays with me for a long time.

Conflict in paradise

This morning I could scratch out the route in the frost on the saddle for the first time. The Himalayas are really upon us now. We ride on small rugged roads to Patnitop, one of the top destinations for Indian honeymooners. In a country where the air vibrates from the heat, snow is the most beautiful status symbol. People sometimes travel up to 3,000 kilometres (2,000 miles) to take a photo with cold feet. For many, happiness is located in the chilliest part of India. The dream of owning a Bullet and driving it to the Himalayas is persistent too. It's the iconic motorbike ridden by heroes and villains in Indian movies. And these days, these dream machines are no longer the preserve of men.

The closer we get to Kashmir, the more humid the air gets, and the more tense the atmosphere becomes. Since the 80s, the valley has been a theatre for armed conflict. In 2002 there was fear of a nuclear war; in 2014 20,000 people fled the area because of conflict. The garden of the Himalayas is the gem that Pakistan, India and China all have eyes for. Among the fertile acres and rice fields, soldiers enjoy naps during the downtime.

We have to ride through the Jawahar Tunnel to the green valley of Srinagar in separate groups. Kashmiris live on the greener grass on the other side and rebels there have pursued independence for decades. In the summer capital Srinagar, we leave our Enfields behind on the banks of the Dal Lake. Gondolas take us to one

Buddhist monks in the making, Zanskar.

of the hundreds of wooden houseboats that form a floating suburb, which are an impressive vestige of colonial times. When the Maharaja forbade the British from possessing land, the elite built their own idyll on the water. Despite the tensions in the valley, the Venice of the north is an important tourist attraction, but mostly for rich Indians, it seems, as foreigners are scared off. Lily pickers, fishermen and boats form a labyrinth on fresh water. It's unbelievable that a war zone can be this idyllic.

On the road

In the following weeks we cruise across the Himalayas. Before we climb the 3,530-metre (11,575-foot) Zoji La Pass, we ride through Drass, the coldest place in India. In the winter it easily hits minus 23 °C here. Yet we still put factor 50 on our noses. My fingers and toes tingle with the cold. When a Tata Truck misses me by a whisker, I'm warm again for at least half an hour.

The road to Kargil is a display of power from road construction workers and army convoys. The fact that the road is in such a good condition is the result of the military game of chess being played here. Their ode to the tarmac is unfortunately one with a fake undertone. The road construction workers battle bravely against the biting wind and obstacles in the form of landslides, melted ice and oppression. Here and there a vat with tar smoulders away over a wooden fire. Later they will mix the tar with stones in wheelbarrows, and pour the boiling mixture over the road. It's hard work, but there are no pessimists. Lesson 2 in humility.

After Kargil we climb up to the really high mountains of the Himalayas. The landscape gets wider, the route unpaved and it's a miracle that the 500cc motorbikes keep going. I am

Into the mountains, on the way to Leh and Khardung La.

not a group person, and prefer to swing on the tail of the flock. Luckily we give each other the freedom to disappear now and then, though no one is allowed to get more than a half-day ahead, because the advantages of being in a group are undeniable when something goes wrong. Flat tyres and gesticulating border police are easier to manage with an extra pair of hands.

The highest pass

Off-road lessons teach you not to cling to the handlebar, but to give your motorbike freedom when it hits a pothole. Be light on your knees, look at the horizon and ride a rodeo. Yeehah! The roar of the engines and rocks flying about barely drown out our chanting. We bounce like cowboys through rugged passes and the Indus Valley. I imagine a motonaut, cruising across the most beautiful moon imaginable. We cross passes of 3,760, 4,147 and 4,880 metres (12,336, 13,606 and 16,010 feet), guiding our Bullets over powdery sand and potholes, through a snowstorm and crossing ice-cold rivers. Days fly by. The road to Lamayuru, Leh and Khardung La is without doubt the most beautiful that I have ever ridden on.

We stay overnight in a tent camp along the salt lake Tso Kar, which at 4,500 metres (14,860 feet)) is the highest sleeping spot of the trip. Most people have a headache because of the altitude – or is it the rum? Big thoughts and panoramas need big glasses.

Crash course

Ultimately the Rohtang Pass turns out to be the biggest challenge, although it is not the highest. Hairpin bends and a bad road surface mean the climb takes an incredibly long time. The road blends into the background. When we are almost at the top we can again enjoy a bit of tarmac. In front of us a beautiful panorama opens up. When I ride through the umpteenth shallow-looking puddle, the slush splashes to the sides. At 2,168 kilometres (1,347 miles), a gigantic pothole appears: 30 centimetres (12 inches) deep and over 50 centimetres (20 inches) in diameter, according to the estimates afterwards. I am riding at barely 50 km/h (30 mph), but cannot get out of the way. Crap. I hold the handlebar tight, take the first shock and feel my back wheel mercilessly disappear into the pothole. Not a moment later Bullet and I are catapulted into the air. We flip over in one, and I break his fall. What follows is a blur.

I ask if someone can pull the motorbike off my leg. The engine is still on, and the prayer flags on my handlebar are flapping at half-mast. One of the bikers walks away looking pale. When I try to get myself up the pain tears through me. My upper arm is completely broken, just under the safety pads that managed to protect my shoulder. I will only find out a week later that my fibula is also broken. I groan. Noises fade away, but the Himalayas stand tall. I don't really register that this means the end of my trip. My clothes are sliced with my penknife. My mind wanders off.

Our guide has no phone reception, because satellite telephones are forbidden here. After a lot of persuasion, a rickety ambulance that is on its way to the north side does a

A Sikh takes a holy dip in front of the golden temple of Harmandir Sahib, the Mecca of Sikhism in the heart of Amritsar.

Sometimes I am dubbed the moaning Titan. Or the one that fell from the Himalayas.

Balloon seller. *(left)*
Brand-new scar: a route marker for life. *(right)*
Traffic jams at a not-so-lonely height. *(following pages)*

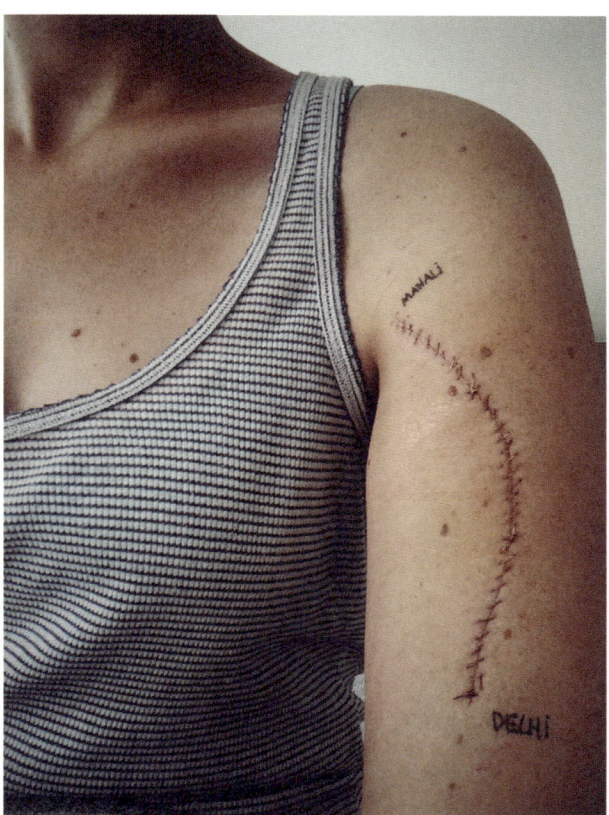

U-turn. There is no doctor and no medicine. But anything is better than nothing now. Luckily no one tells me that Manali is over five hours away. I end up on a stretcher and get some cardboard as a splint. Ultimately it will take us six days to get to Delhi and then Belgium. Panoramic views make way for ceilings of sick bays, hotel rooms and ambulances. It's 700 bumpy kilometres (435 bumpy miles) over land. A second road trip, with sirens. Thanks to my first accident in 18 years of riding a motorbike, I fly first class for the first time in my life. After six months of wheelchair driving and rehabilitation I'm back on the road. The scars are a living memory. Sometimes I am dubbed the moaning Titan. Or the one that fell from the Himalayas. //

LOCATION

Himalayas
India, South Asia

CURIOSITY

The Royal Enfield Classic Bullet motorcycle wove itself into the hearts and minds of the Indian population, not the least because of its signature thump and retro design themes. Owning one and driving it up north through the Himalayas has been a dream for many. A high-altitude odyssey across generations.

ADVENTURE TIME

The Ladakh region is almost always dry, but even more so during the summer months (June–September).

TRAVEL TIP

Prepare for all weather conditions. Temperatures in Leh can reach -35 °C and up to 35 °C in the summer. High altitudes and deserts can result in icy conditions, and snow on the high peaks lasts until early July.

BEFORE YOU GO

The extensive drive requires some experience. Take some off-road lessons to prepare yourself for this physical challenge.

IN A
DIFFERENT
LIGHT

The Art Islands

Strolling around Naoshima and Teshima

For anyone who hates having to choose, Japan is a country where you really can have it all. Nowhere can you dash faster from the futuristic to the historic, the serene to the eclectic, the artificial to the pure. In Japan, this more often results in an interplay than a clash, something that's best experienced on its art islands.

Naoshima and Teshima are a universe of their own. Under the guidance and inspiration of Fukutake, these sleepy islands in the Seto Inland Sea developed as a place where art, nature and architecture merge. アート巡礼、 an art pilgrimage. Sunken in the hills. Rooted in a desire for connection.

It takes a bullet train to Okayama, a slow train to Uno and a ferry to Naoshima's Miyanoura Port to reach the islands. Hours during which I peel off the hustle and bustle of Tokyo one layer at a time. And that winding down turns out to be intentional. Japanese billionaire Soichiro Fukutake hoped to create a country within a country, nearly two hours from the nearest town. An open treasure trove, tucked away between more than 3,000 islands, almost all uninhabited. 'A map of the world that does not include Utopia is not worth even glancing at', Oscar Wilde wrote, 'for it leaves out the one country at which Humanity is always landing.'

After visiting Naoshima in 1985, Fukutake was moved by the way local fishermen still formed a close-knit community, their lives dependent on the sea. A symbiosis that visibly crumbled as the islands were sacrificed for the modernization of post-war Japan. Companies illegally dumped tonnes of industrial waste on Teshima, and on Naoshima copper refineries emitted unfiltered sulphur dioxide. Young people left in search of a better life in the big city.

Touched by their fate, Fukatake changed his plans. He decided to bring them back to life, and develop his art project not in the metropolis of Tokyo, but here. 'The city is the destructive centre of chaos, money and stress. There's too much entertainment, too many people and too much stuff. Art and culture should not be confined between white walls. The natural world is the only place that can do justice to art.'

He changed the name of his family empire to Benesse (well-being in Latin), hired architect Tadao Ando to create the Benesse House Museum, Chichu Art Museum and other nature-first galleries, and turned them into a condemnation of advancing industrialization.

Art for the heart

When the ferry drifts into Naoshima harbour, I immediately come across a work of art. The large red pumpkin by the Japanese pop art artist Yayoi Kusama is a clear statement: rain, sliding about, daydreaming – such works are open to everything.

'Art must exist amidst nature' is the slogan of the project Fukuyama has been working on for over 30 years. A secret ploy to teach us to look at both forces with fresh eyes. Most works of art are therefore only accessible on foot, along country roads dotted with

The Teardrop at Teshima's Art Museum: a concrete cocoon between the rice fields.

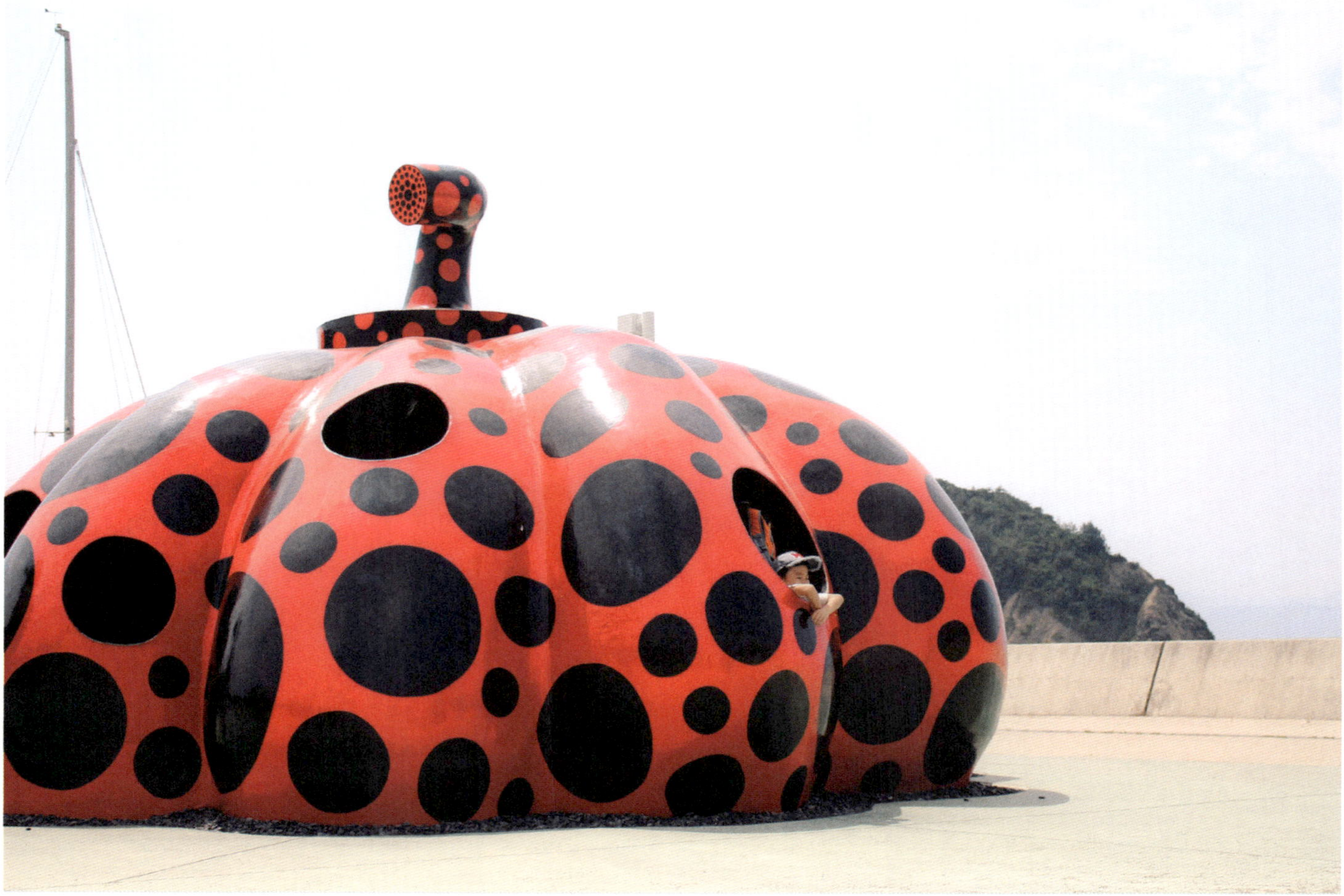

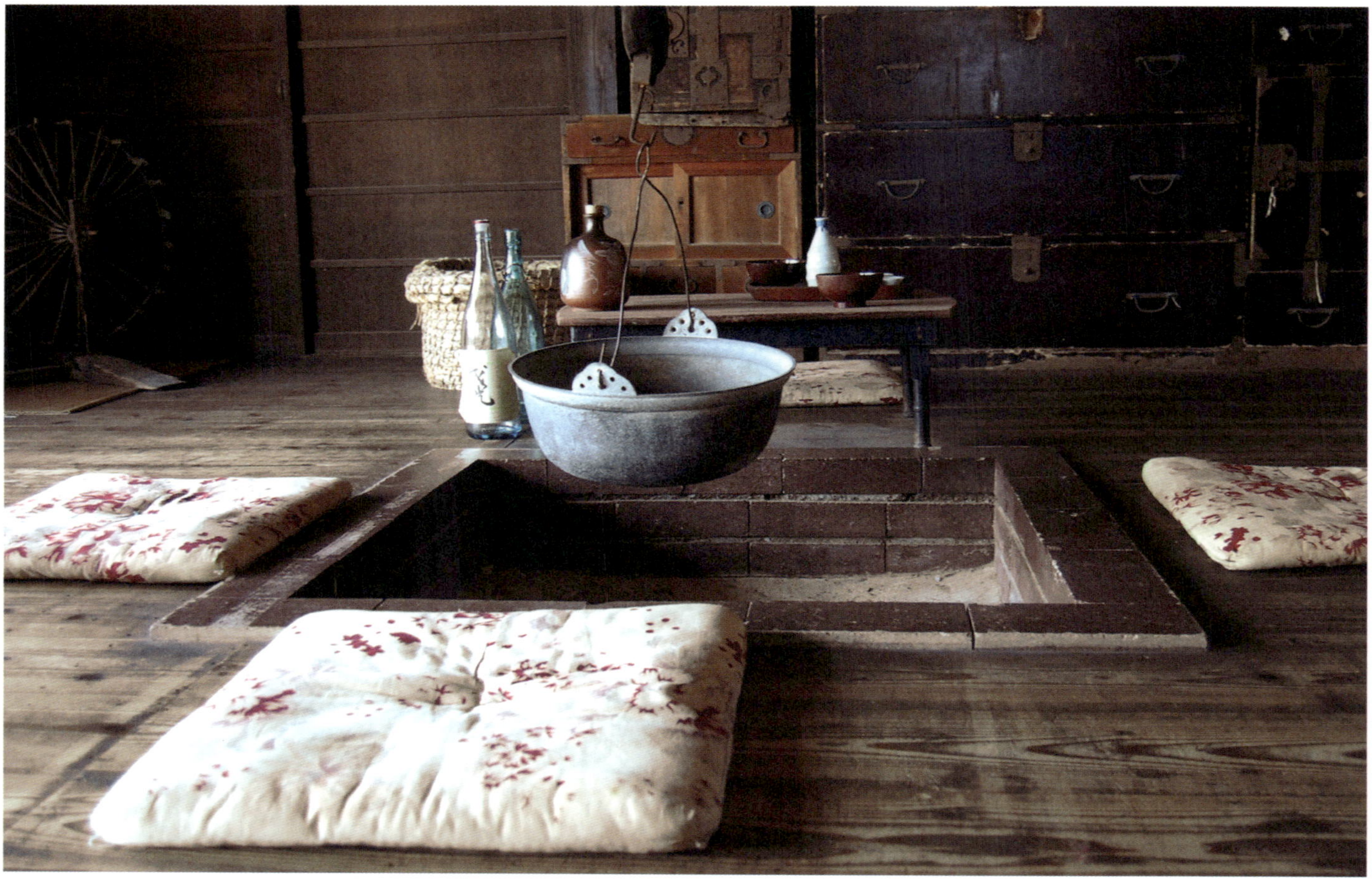

blossoming trees and chirping birds. Tadao Ando's structures resemble temples, stripped to a bare simplicity, influenced by Zen Buddhism and the wabi-sabi aesthetic. No wonder some compare a trip to and around the island with an art pilgrimage.

I walk to the other side of the island and discover the first works along the coastline, in the narrow streets and the forest. I spend the night in a caravan on the beach, with a view of the gigantic, this time yellow, pumpkin by Kusama. A surreal beacon sticking out of the sea. Little did we know it was destined to be swept off the shore during a fierce typhoon. On the hillside, guests will soon wake up in Benesse House amidst iconic works by Andy Warhol, David Hockney, Jean-Michel Basquiat, Richard Long and Keith Haring, in the Museum, Oval, Park or Beach Room. After midnight, in slippers and a bathrobe, is undoubtedly the most intimate time to experience the works.

In *I Love Yu sentō*, a public bathhouse and artwork, the artist Otake Shinro has also turned the museum inside out: inside became outside, public became private. The exterior is garish, with neon lights, colourful tiles and mirrors. Inside you're bathed in a scrapbook of memories.

The Chichu Art Museum was built as an ode around its iconic works, such as Monet's *The Water Lily Pond*. I descend into the concrete corridors, deeper and deeper into the hill. An underground passage to get me in the right mood. Footsteps echo. Other visitors become silhouettes. The outside seeps through cracks

The red pumpkin by the Japanese artist Yayoi Kusama. *(top)* Taking a look inside the Art House Project: a collection of abandoned houses and workshops transformed into art installations. *(bottom)*

and holes. The guards move through the rooms in white lab coats. The Monet Room has no corners. A Walter De Maria installation of 27 gold sculptures resembles a science-fiction scene. Fragments of whispered Japanese make the experience even quirkier. All my senses seem heightened. As we come out, everyone remains subdued. Moved by it all, someone points to the shadow of a swaying branch on a rock.

Teardrop

Viewing art here is much more about touching it. There's no red ribbon or route. Everyone is free to set their own pace as they explore the islands and their artworks, which have captivating names such as *Needle Factory, Storm House, Sea of Time, the Missing Post Office, La Forêt des Murmures* and *the Life Garden*.

Personally, I feel the most moved on neighbouring island Teshima, which is even smaller and more intimate. In *Les Archives du Cœur*, by Christian Boltanski, we wander through three rooms: the installation in the Heart Room, a recording studio where you can add your own heartbeat to the artwork, and a listening room.

Then we walk into the hills to reach the teardrop-shaped Teshima Art Museum. The vision of artist Rei Naito and architect Ryue Nishizawa, it's a sloping concrete cocoon between the rice fields. Just like in a temple, I have to wear slippers. Once inside, photography, writing, talking – in short, anything that's a distraction – is banned. Drops of water bubble up from the ground, then move along almost invisible paths, only to disappear a little further on. Every season or weather change creates a different view of this world, changes the experience, and therefore also the work. I lie there on the ground glistening for more than half an hour. The spell works. Life imitating art. Ordinary life elevated to art. //

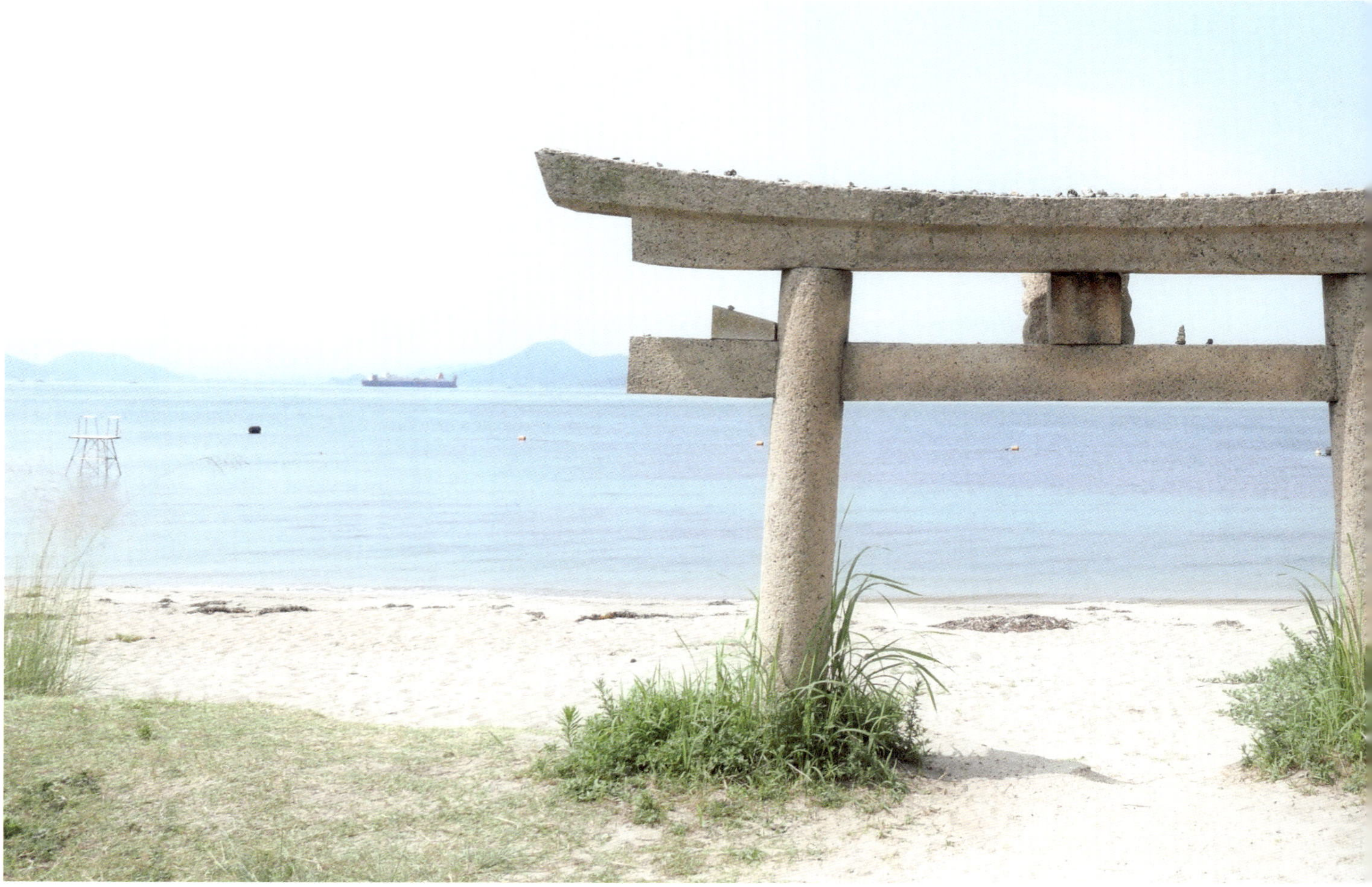

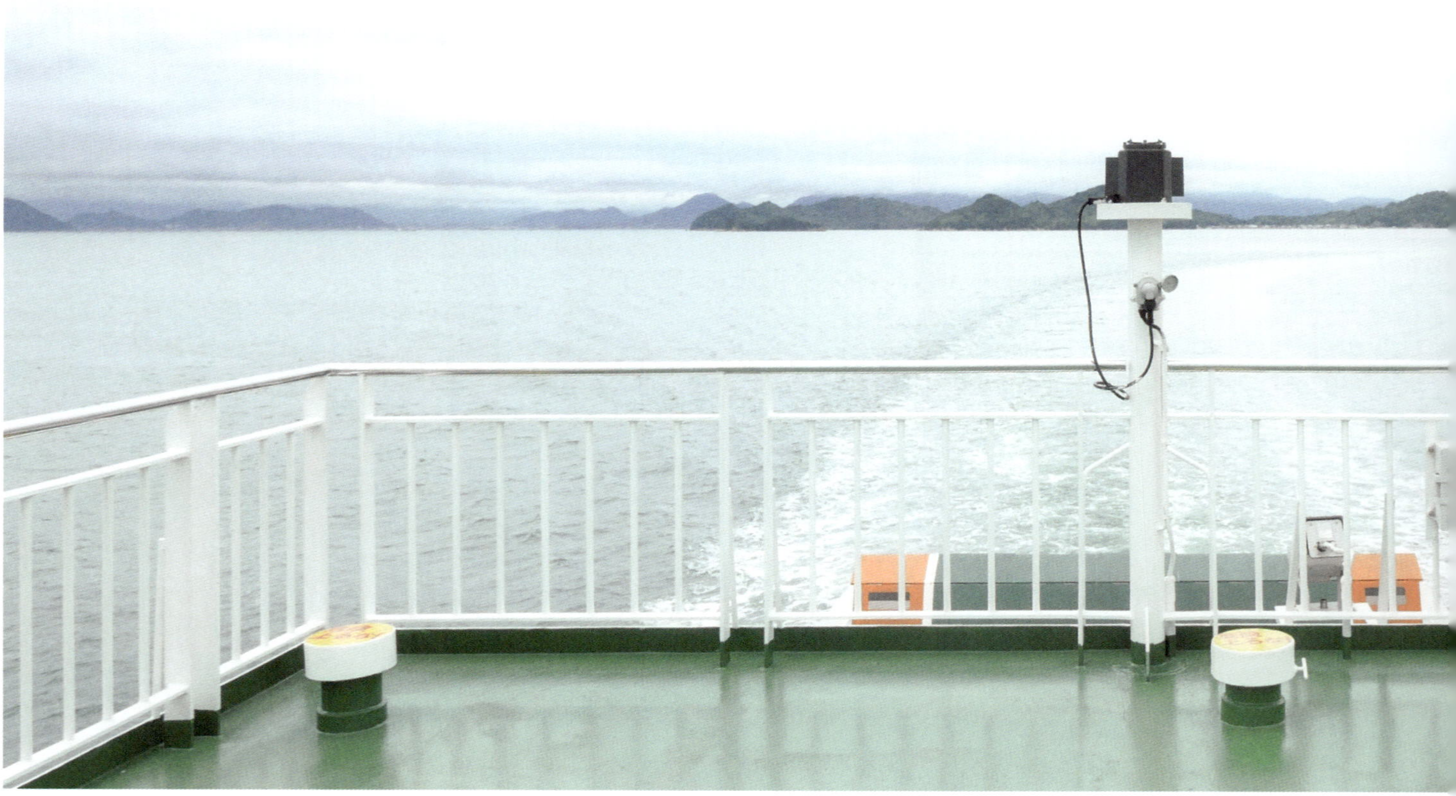

No wonder some compare a trip to and around the island with an art pilgrimage.

There is also room for more traditional constructions, like this Shinto *torii*: a gateway from the everyday to the sacred. *(top)* The colourful deck of the ferry to Naoshima's Miyanoura Port. *(bottom)*

LOCATION

Naoshima and Teshima
Japan, East Asia

CURIOSITY

Monets, underground crypts and stunning Tadao Ando architecture are hidden on a handful of remote Japanese islands. Thanks to a private benefactor, these once forgotten fisherman's islands have been transformed into an art mecca that navigates the delicate balance of maintaining tradition and accepting modernity.

ADVENTURE TIME

Warm and pleasant daytime temperatures, coupled with the right amount of dampness in the air, makes May the best time to visit the islands.

TRAVEL TIP

Book a night in Benesse House on Naoshima. It functions as both hotel and art gallery, letting you fulfil your wildest dreams of having a sleepover with Andy Warhol.

BEFORE YOU GO

Before planning your trip, be sure to consult the Benesse Art Site Calendar.

A glimpse inside Tadao
Ando's Benesse House
Museum. *(top) One Hundred
Live and Die*, Bruce
Nauman's brutally honest
1984 neon installation.
(right)

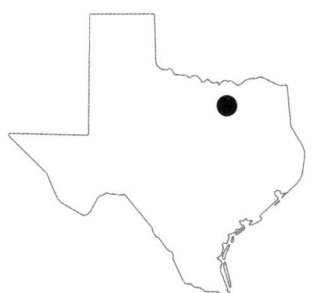

The New Wild West

Taming the clichés of the Lone Star State

A shot rings out. A saloon door clatters. In the distance, a clump of tumbleweed rolls on by. Contemporary Texas rises above the clichés, but also does everything it can to keep alive the legends that have shaped its young history. Welcome to the enigmatic Wild West.

My nose is glued to the plane window as we fly over Greenland. About 9,750 metres (32,000 feet) below us, the coldest point on our planet appears out of nowhere. Pristine in its shades of white, violet and pink. We get a helicopter view of the frozen landscape, which from up here looks more fragile than unforgiving. Down below, the polar ice caps crack, tear and melt.

I keep staring at the same spot as a little plane on the screen in front of me draws a yellow arc from Reykjavík over Newfoundland and Labrador to Chicago and then Dallas. This little interlude has made me even more curious about Texas. A state we've heard so much about, yet also not enough. About its turbulent past, for example. Or its sustainable futures, given the cultural importance of barbecuing.

On the trail of the Barrow Gang

From north to south, the 2695,666 km² (270,000 square miles) of Texas span a distance of 1,280 kilometres (800 miles). This makes it the second largest state in the United States, after Alaska. My factual knowledge of the region doesn't extend much beyond the clichés: those well-known stories about the Old Wild West. The eccentric characters who managed to immortalize their short lives have been capturing our collective imagination for

Cadillac Ranch RV Park, Hope Road, Amarillo.

years. The success of the PlayStation game Red Dead Redemption 2 and the Westworld series are living proof of that.

It certainly isn't a new hype. When we were growing up, my brother and I spent hours playing Cowboys and Indians, devouring the Bluecoats and Lucky Luke stories, while hearing the Dallas theme tune blaring out of the living room. It is only now, as an adult, that I realize how one-sided those stories were. What are things like today? And what characterizes modern Texans?

As soon as I arrive in Dallas, I realize they take the slogan 'Everything's Bigger in Texas' pretty seriously. From the queen-size hotel bed and the gallon of coffee to the adjectives they use. It takes a bit of getting used to, especially for a modest Belgian like me.

For the next few weeks, my playground will be in West Texas. It doesn't get much more Wild West than this. The place where, not so long ago, Native Americans, cowboys, bandits, outlaws and bisons roamed the endless prairie. Where the streets were just a strip of gambling houses and saloons. The birthplace of rock 'n' roll legend Buddy Holly, and the land 'most wanteds' such as Jesse James once called home, if only for a little while.

Even the most notorious outlaws in American history, Bonnie Parker and Clyde Barrow, made this their base. Right before their adventure began, Bonnie was working in a Dallas café. Once that adventure was over, this was where Clyde was buried. Nearly 100 years later, the story of these young bank robbers and murderers still captures the imagination, as do the places they once went. I too am on their trail, heading north-west.

Cattle & cotton

I leave Dallas and pick up the RV that I'll be sharing with a few unknown travel companions for the next week. As far as both the vehicle and the state are concerned, I'm entering uncharted territory. Having never seen a 5-metre-long (16-foot-long) recreational vehicle before, I'm in awe of the space and gadgets it offers. For a wild camper like me who normally doesn't even let a single pair of extra socks through beacause of her ruthless packing criteria, this style of travelling is a whole new world.

Once we've driven out of the Dallas suburbs, the view stays the same for hours. Driving becomes an exercise in mindfulness, accompanied by the humming of the engine and the strumming of the country radio stations. The landscape is flat and the road only seems to head in one direction – the horizon. Here and there, the monotony of the prairie is broken up by a village or a signpost and a turn-off, the only indication there might be something else around.

We spend our first night in Jellystone Park in Wichita Falls, the first big city, about 240 kilometres (150 miles) north-west of Dallas. The engine of the RV is still ticking away when the local sheriff greets us with a firm handshake. Tonight, he's trading his star for an apron and his revolver for some tongs to sear a dozen 2kg (70-ounce) steaks on a wood

fire grill. A bit later, he produces a peach cobbler too – the traditional dessert that has kept many a cowboy warm on the prairie. He is brimming with pride as knives glide through the hunk of meat. To him, steak size is still the best way to measure Gross National Happiness.

No surprise then that livestock is still one of the state's biggest riches. Aside from oil, gas, grass and cotton, the dry soil here produces very little that is useful. Promoting meat remains a matter of honour, for the time being. Today, the sheriff will make his first acquaintance with a vegetarian, respectfully and with just a touch of pity.

Yet the vegan lifestyle is growing in popularity here. You might not expect it in the land of barbecue and Tex-Mex, but today you'll find more than 1,700 vegan-friendly restaurants, veggie fairs and other pop-up markets. 'Black veganism' is also gaining a following. For some, the choice is a silent form of protest.

Campers & canyons

The next day we drive further west. This is the way to Amarillo. With a cotton field here, a lone watermill there and then nothing again for hours. On autopilot, the RV chugs its way towards the Texas Panhandle. Only the odd bump in the road can break our meditative flow. The crockery responds in Morse code tones.

At the very top of the Lone Star State, the landscape

Palo Duro Park. Stories of Native Americans add colour to the canyons.

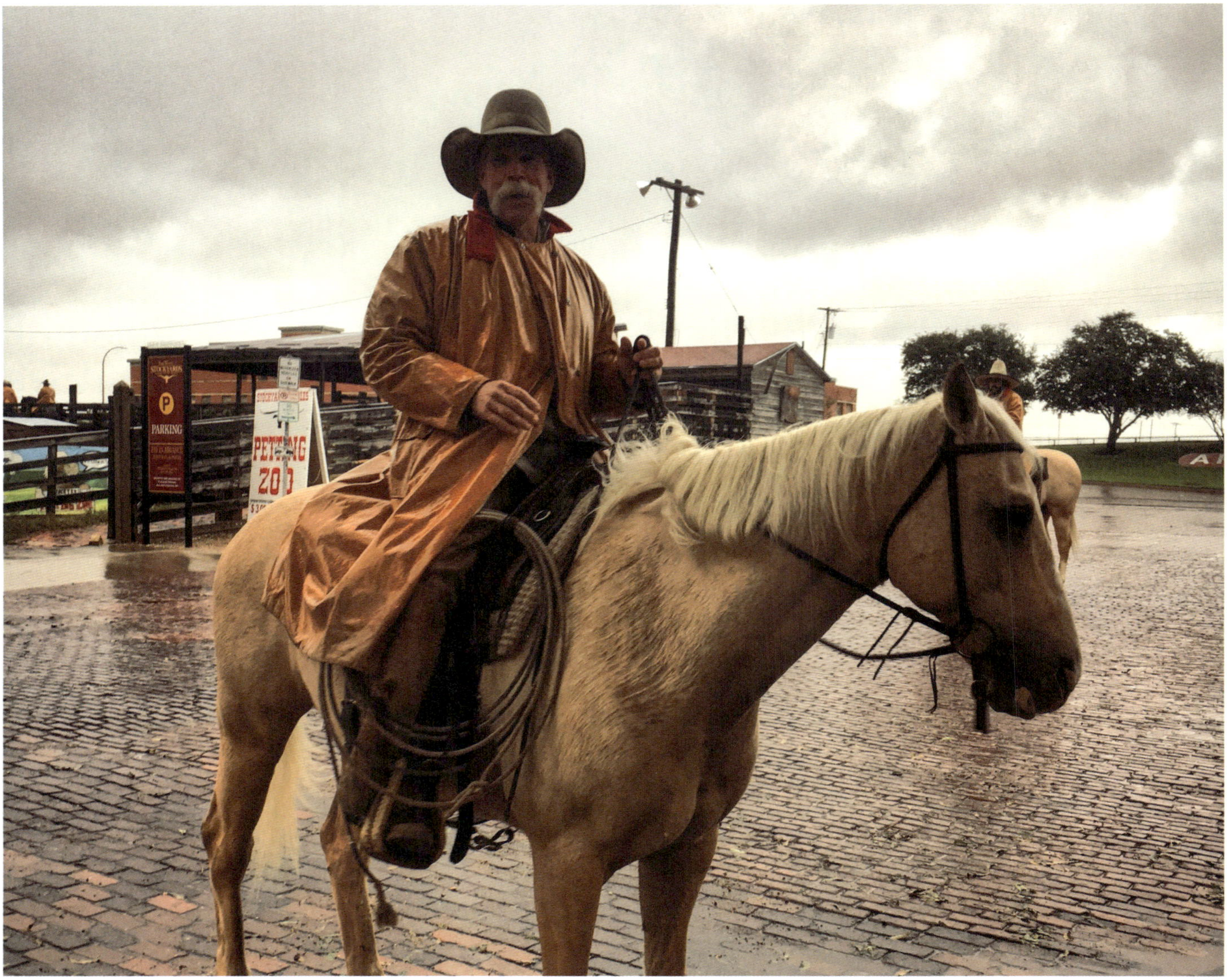
STOCKYARDS
P
PARKING

DROVER
TEXAS
JWV 0645
The Lone Star State

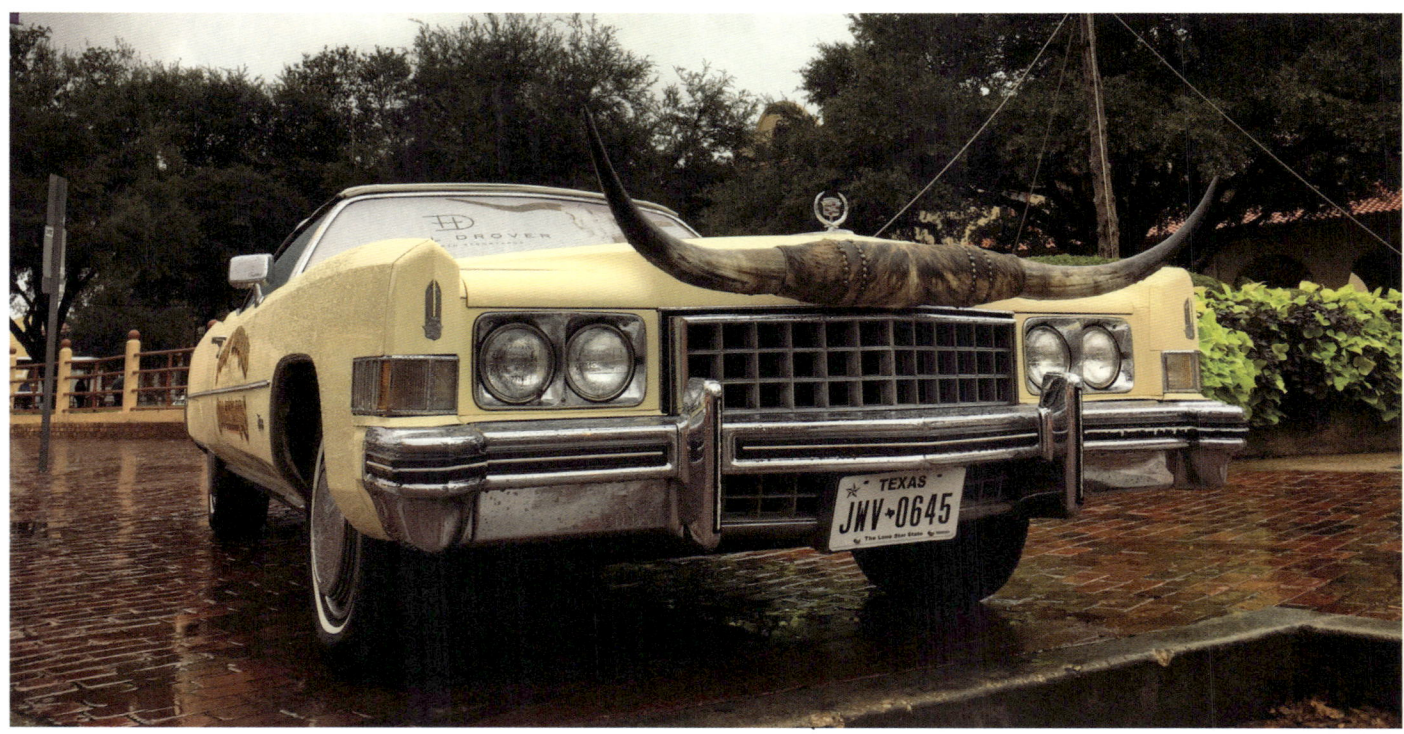

Driving becomes an exercise in mindfulness, accompanied by the humming of the engine and the strumming of the country radio stations.

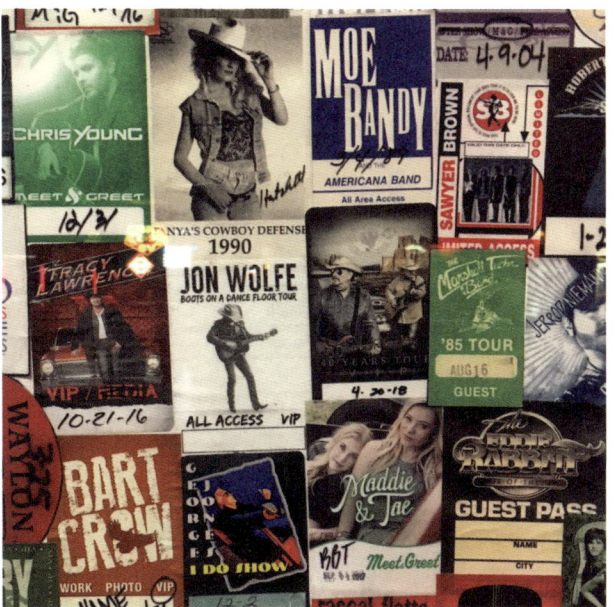

suddenly plunges into the depths, between the High Plains and the lower plains of Central Texas. At 15,313 acres (6,170 hectares), Caprock Canyons State Park is one of the state's largest parks. The narrow ravines make the rock faces more accessible and the colours more vibrant. The cliffs soar to an immense height, making a natural shelter for the pathways and streams. You can spend hours hiking, mountain biking or horseback riding on its 40 kilometres (25 miles) of trails.

The road to our campsite slips right through the layered canyon walls, into the valley of the Little Red River. It's hard to imagine a small stream carving such a deep gorge into the landscape. Easier to picture are teepees dotted along the wide bend in the river. It was in these canyons that Apache and Comanche tribes used to shelter from the bad weather and hunt for the wild bison, the animal that provided them with everything they needed. During the so-called 'buffalo jump', they would chase entire herds into a ravine. Without any compassion, but with respect for the ecosystem. The buffalo is a symbol of life, health and well-being for Native Americans.

After the unbridled hunting of the animal by Buffalo Bill and his associates, the bison was on the brink of extinction. There's no doubt about it: this mythical figure was a serial killer, who was responsible for the slaughter of as many as 60 million bisons in the hope of forcing the Native Americans into submission.

This colourful landscape was also once part of Charles Goodnight's legendary JA Ranch. He was the one who brought tubs of bison seed and a herd of bison to the park, pressured into it by his wife who was determined to protect the animals. This herd would later

No shortage of cowboys and curiosities in Fort Worth.

become the official, protected Texas state bison herd. A symbolic link to the past, with everything that entails. The return of the bison goes hand in hand with a revival of the Native American identity, which is deeply connected to nature. A story of recovery and rewilding.

We set up the camper in the middle of the canyon and make it coyote- and bison-proof. Because this wouldn't be America if you weren't allowed to drive your RV not only to the edge of the canyon, but also right to the bottom. It cools down surprisingly quickly. During the day it's sweltering, even in October, but at night our breath turns into wispy clouds. We're treated to a galaxy that lights up like a diamond-studded pincushion. The next morning I see my first halo around the moon.

The state is so big you start to wonder where its 27 million inhabitants are hiding.

Buddy Holly

After a night in the wilderness, we drive back to civilization via Lubbock in the direction of Abilene. 'Civilization' is perhaps a stretch. The state is so big you start to wonder where its 27 million inhabitants are hiding. The streets in the towns and villages are suspiciously quiet. All you see here are other campers and cars moving, often with darkened windows. That desolate feeling only adds to the 'Texas mystique'.

Lubbock's claim to fame is as the birthplace of rock 'n' roll legend Buddy Holly. The singer-songwriter's talent and the story of his stormy career accompanied his songs all around the world. We visit the museum, his family home and grave, where he was laid to rest as an eternal 22-year-old under plastic flowers and guitar picks. At the Buddy Holly Center, fans of Peggy Sue's spiritual father can admire collector's items such as his black horn-rimmed glasses and his Fender Stratocaster.

During our last stage the city of Abilene prides itself on bizarre titles like 'city with the largest number of churches per resident', 'ninth most windy city of the US' and 'Storybook Capital of American sculpture and activities'.

The Frontier Texas museum is also worth a visit. Especially if you're keen to hear the (full) story of the cattle drives, bisons, Native American tribes and camps. It's good to be

Texas, land of roadstops *(left)* and cowtown shops. *(right)*

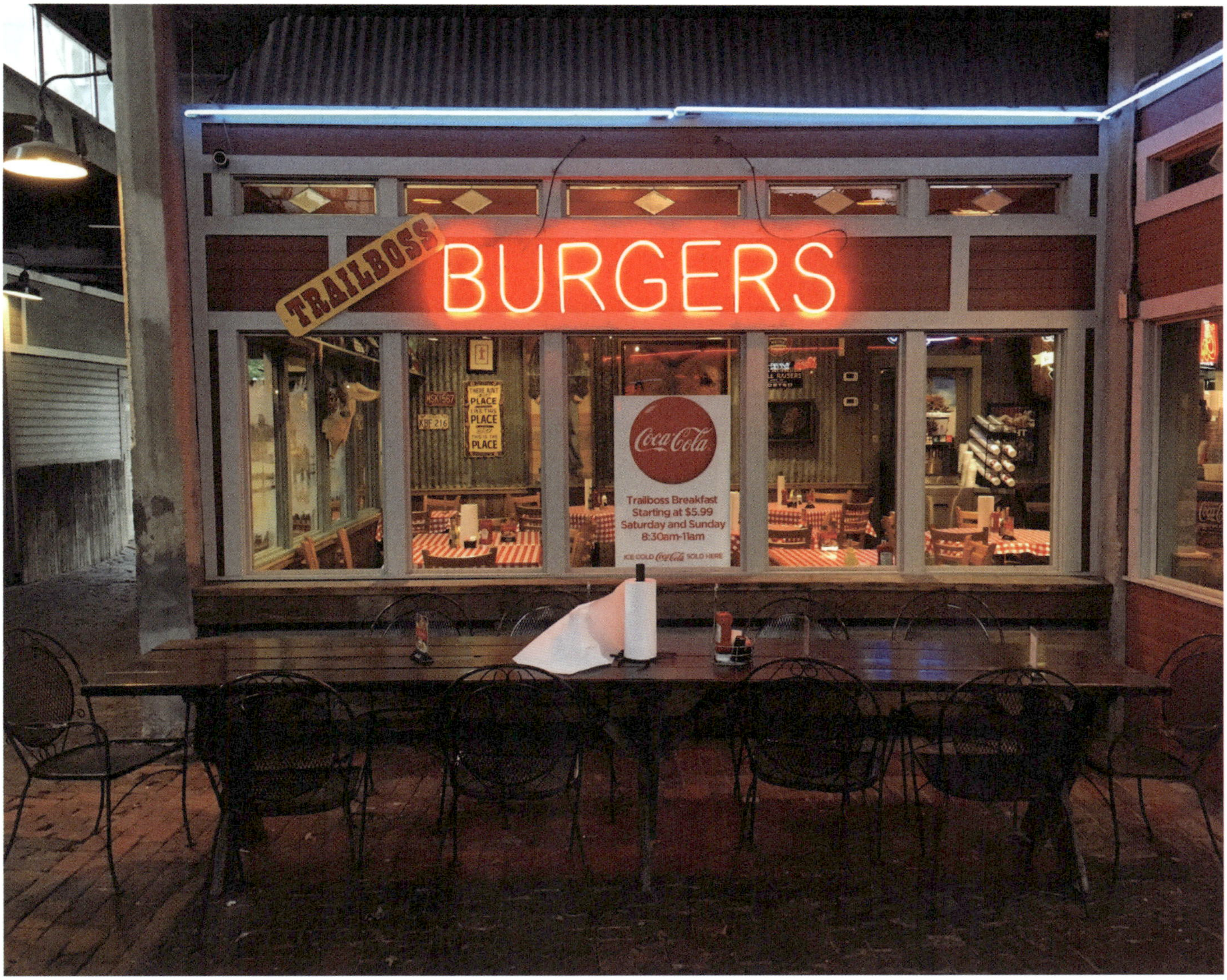
TRAILBOSS
BURGERS
Coca-Cola
Trailboss Breakfast
Starting at $5.99
Saturday and Sunday
8:30am-11am

Shiner
IT'S A
TEXAS THANG
Y'ALL

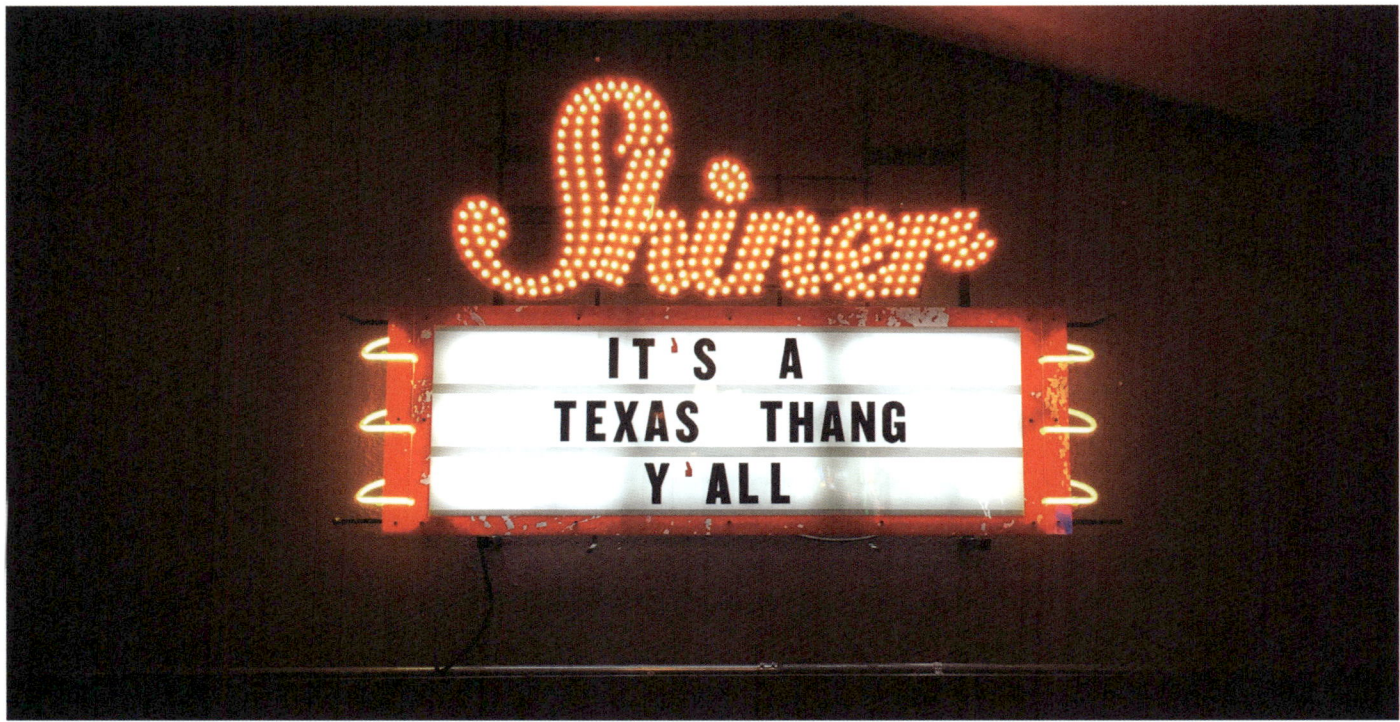

aware that the visitor centres in the parks are often still toe-curlingly inaccurate. This museum has found the right technology and tone to explore a short but very sensitive period of history in a captivating way.

Old forts & Fords

Our trip through the best of the west ends in Fort Worth, aka Cowtown. For many visitors, this is the highlight of the trip. This is where bandits, cowboys, outlaws, Native Americans and sheriffs used to fight each other by day, then pull up a saloon stool together by night.

Fort Worth was founded in 1849 as an army post to protect settlers from attacks by 'enemy Indians'. Later it became a stopover point for cowboys on the cattle drive trail to the Wild West. Thousands of wild longhorns were led through the streets here to the stockyards. With the arrival of the railways and the discovery of oil, the city really started to boom.

Living legends Bonnie and Clyde also stayed here, to pay a 'visit' to the banks. At the Stockyards Hotel you can sleep with Bonnie's pistol above your bed. It was the US dollar bills that these bandits were after. And even today, those notes are still freshly pressed and printed here, at the US Bureau of Engraving and Printing in north Fort Worth. A visit to 'the Money Factory' won't cost you a dime.

Other major attractions include the stockyards, which cowboys parade through daily, the TX whiskey distillery and Billy Bob's, the largest honky tonk in the US. Presidents, cowboys and stars from all over Texas and the United States shop at M.L. Leddy's leather store. They're searching for a bit of hero

Fort Worth, a crossroads of clichés and new cultures.

The rich mix of cultures and migrants defines the vibrant character of the Lone Star State.

status and the iconic pair of handmade boots, available custom-made in all colours and sizes. Ranging from about $1,000 to $14,000 per pair, so you'll need to return a year later to pick them up.

The best of the West

All in all, the Lone Star State managed to surprise me. It is the rich mix of cultures and migrants that defines the vibrant character of this state. You hear it in the music, taste it in the food and experience it in the warm-heartedness of the people.

Dallas sits right on a tangible line between the nostalgic Old West and the more modern east. To the left, Fort Worth beckons with its soundtrack of Buddy Holly and country music; to the right, vegan-friendly Austin grooves along to Khruanbing and Cigarettes After Sex. The state is gradually throwing off its old skin, revealing a colourful and layered story that is best explored slowly. *Y'all have a great time, folks.* //

Leddy's leather boot collection. *(previous pages)* **Billy Bob's, the largest** *honky tonk* **in the US.** *(left)*

LOCATION

Lone Star State
Texas, USA

CURIOSITY

Texas isn't only the land of cowboys and cacti, a place of barren landscapes where cattle rustlers still roam. Uncover the different layers of the Lone Star State and tame the clichés of the Old Wild West.

ADVENTURE TIME

Fewer tourists flock to Texas in April and September, but with slightly cooler temperatures than summer, these are both great months to make the most of outdoor activities.

TRAVEL TIP

Visit the Texas state bison herd, and uncover a journey from near extinction to celebration. Plans are to rewild the plains, by bringing back buffalo and prairie dogs to restore ecological balance.

BEFORE YOU GO

Go full Texan, and expand your vocabulary with a lot of adjectives and superlatives.

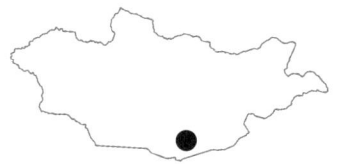

13

Modern-day Nomads

Eating dust in the Gobi Desert

Places without a single well-trodden path do exist. Mongolia has just three million inhabitants, more than half of whom live in the capital. Around the city, panoramic views stretch for hundreds of miles of nothing but vastness.

I cross the country without fences, walls or barriers, not on horseback, as is tradition, but still with the necessary horsepower. All I have to do is let go – of both my control and the clutch – and ride, ride, ride. A hypnotic 2,392-kilometre (1,486-mile) trip through the Gobi Desert and the 'Land of the Eternal Blue Sky'.

The Eastern Bloc and a first stretch of Russia are now behind us. Over the past few weeks, my two fellow travellers and I have covered the first 8,000 kilometres (5,000 miles) of our cross-continental motorcycle ride from Brussels to New York – the first leg of our road trip through Europe, the former Soviet Union and America. With temperatures fluctuating from around 8 °C in the Carpathians to 51 °C on the hot Russian tarmac, it was an endurance test for both suspension and resilience, just months before a global pandemic and war that would forever change the lives of those whose paths we crossed.

Russian Tibet

After 20 days of flat land, a ridge finally beckons in the distance. The Altai Mountains, in the Russian republic of the same name, feel like a country in their own right. We haven't yet officially crossed a geographical border or time zone, but after riding for several hours everything suddenly feels different. This 'Russian Tibet' stands at the crossroads of various states and cultures, and is the place where the Siberian taiga, the steppes of Kazakhstan

Comparing horsepower in the Ulaan Taiga Mountains.

and the semi-deserts of Mongolia meet. You can sense it, too. The people, buildings and customs suddenly look very oriental. Meanwhile the forests, log cabins with deer heads and bearskins, glacial rivers and rafting and fishing activities have a very Alaskan or Siberian feel. The only indications we're still in Russia are the Lenin statues on the village squares and a billboard of Putin propaganda.

We cross the mountains towards Mongolia, sharing the road with a few Siberian truckers and rust buckets. It gets noticeably quieter the further east we ride. After all, there's not a lot of reason to come here. As we ride to Tashanta and the border with Mongolia, we come across giant signs with weird pictures of marmots and ticks. Pretty funny, we think: those animals must really have gotten up to no good to get that level of attention. Only later that evening, when we finally manage to translate the signs, do we realize they were warning of a national outbreak of the plague. Also known as 'The Black Death'. A sense of otherworldliness overwhelms us. We decide to ride on anyway, and keep our wits about us. It's an unprecedented reality check in a world before the coronavirus.

Least populated country in the world

After eight long hours at the border, we ride into Mongolia just in time: tomorrow morning our visa expires and we're officially no longer allowed to enter the country. Phew. Once across the border, we enter a void. Time for concentration, contemplation and a seemingly endless horizon.

After all, you never know exactly how long it'll take you to cover the next 402 kilometres (250 miles). Two hours or two days. Off-road routes fan out into a handful of tracks that seem to send you randomly into the plains. There's no other option than to choose one and stick with it. All roads lead to Ulaanbaatar! Except for the one going up to the yurt on that faint hilltop... My travel buddies and I only get separated once, but it takes us more than two hours to find each other again. I'm sweating, and not just from the slight panic. With temperatures of up to 40 °C, and not a house, garden or tree in sight, our only respite comes from the shade of our bikes. We're lugging around extra litres of water in our panniers. After 200 to 300 kilometres (150 to 200 miles) or so, we finally spot a village and refuel without hesitation. Coffee, water, petrol and a suspicious-looking plate of crispy chicken.

We ride until we run out of both energy and daylight. Then we set up camp, crawl into our sleeping bags and are left with nothing to count but the stars in the Milky Way. It's so dark here that it's the sky itself that lights up. I battle tiredness just to stargaze a little longer.

These Mongolian steppes have an impressive past. Under Genghis Khan, the country left its mark on world history. The landscape is stunning, but ruthless. We ride in the footsteps of camels, yaks, wild horses, riders, shepherds and their goats. Yet we barely meet a living soul along the way. Just skeletons in miles of nothing. After 250 kilometres (150 miles) of off-road riding, we could kiss the first bit of tarmac or traffic we come across.

Hybrid living

A week of eating dust later, we reach the capital Ulaanbaatar, which unites old and new Mongolia. From a distance, I can see the city emerging. Or something that looks like it. Many Mongolians and expats come here to seek refuge, and the city is growing at lightning speed. More and more artificial city districts are popping up around it, made up of the yurts and dilapidated houses of fortune seekers swapping their nomadic existence for a city view. But for many, the situation seems hopeless. The big city dream that the Korean soaps dangle in front of them only comes true for the lucky few.

We first need to make our way through the smog, which hangs over the city like an ominous vapour. It's not so

much traffic, but industry and the many coal and wood-burning ovens that appear to be the biggest polluters. There's no ring road around the city, so all the traffic has to go straight down the main street. And so we're stuck in traffic, in the heat, for hours. But a strange silence reigns in the city centre, because the modern Mongolian doesn't drive a thoroughbred, but a hybrid.

Earlier we even came across a Toyota Prius in the middle of the desert. They're apparently a real bargain here, thanks to Japan and its strict safety regulations. Lots of Japanese people decide to upgrade their car when its first test is due, and a significant amount of those written-off models are exported to Mongolia. To make traffic more environmentally friendly, the Mongolian government also offers various tax exemptions on hybrid vehicles. So don't be alarmed if, somewhere in this middle of nowhere, you spot a yurt with a satellite dish on the roof of a Prius. //

Even in the middle of the city, the modern Mongolian often opts for the traditional yurt. *(left)* The rugged, rusty foothills of the Altai Mountains. *(right)*

Graveyard for the competition horses of the Naadam: traditional races where mainly the horse is honoured after a victory. *(previous pages)* A glimpse inside Naran Tuul, better known as the Black Market. *(left)* Crossing the Gobi Desert. *(following pages)*

LOCATION

Gobi Desert
Mongolia, East Asia

CURIOSITY

In the mystical steppe and the remote dunes of the Gobi Desert lives one of the world's last surviving nomadic cultures, whose customs pre-date the age of Genghis Khan – although modern nomads seem to be swapping their horses for hybrid cars and motorcycles.

ADVENTURE TIME

The best time to visit Mongolia is between mid-June and late August, which is the summer season, characterized by sunny days and a little rain.

TRAVEL TIP

Spend a night under the stars in the desert and gaze up into the galaxy.

BEFORE YOU GO

Watch the documentary *Rally for Ranger* about a group of adventurers from around the world who band together to deliver new motorcycles to rangers patrolling the Mongolian heartland. Without these precious vehicles, the park rangers working tirelessly to protect ecologically important regions would have to patrol thousands of miles on foot or horseback.

14

The Flyoverland

Uncovering the roots of Route 66

The rugged midsection of the US often gets overlooked. Or rather, flown over. Many people dash through it in about five hours, seamlessly, from metropolis to metropolis. At both ends, the coastal areas have left their mark on the country's identity. But there's more to the in-between space than prairies and motels.

An overland drive from California to New York gives you the time to unravel the many faces of the US. Following the iconic Route 66, we ride right through America's heartland. In search of the Road to Opportunity in all its guises.

Our cross-country ride starts at the Route 66 landmark on Santa Monica Pier. Officially, this sign marks the end of the iconic route, but many decide to start here instead. We too will soon leave behind the cool breeze of the Pacific Ocean. From here we'll ride through the heat of the central states towards New York. After having hugged the coast for 2,012 kilometres (1,250 miles), California is where we pivot inland.

We've spent the last week riding here from the Canadian border, the water always at our side. The 101 Oregon Coast and Pacific Coast Highways guided us through Santa Cruz, Malibu, Santa Barbara and Venice Beach to Los Angeles. A route so deeply ingrained in our pop culture that we could have ridden it almost with our eyes closed. Past boulevards, palm trees, seaside houses with views of coves where dolphins swim and beaches frequented by pelicans and pop stars alike. The land of plenty. In Oregon we spotted whales, and in California the ancient sequoia trees in Redwood and Humboldt National State Parks took our breath away. Between their roots, our little tent vanished into thin air.

Synthetic splendour also stole the show. Rarely have I seen so many Mustangs, Dodges, Maseratis and other flashy motors in such a small area. We marvelled at everyone wanting to be seen on Venice Beach, and even caught a midnight screening of Once Upon a Time in Hollywood at Tarantino's arthouse cinema on Beverly Boulevard. An evening among the greats, just like in the movies. The contrast with the wilderness of the previous weeks was equally cinematic.

Death by Interstate

We skip the Walk of Fame and ride along Historic Route 66, one of the world's most legendary roads, towards Oklahoma. For the next 3,000 kilometres (1,870 miles), the 'Mother Road' guides us through Arizona, New Mexico, a stretch of Texas and Oklahoma. Nowhere captures the spirit of the American dream like this. Opened in 1926 and officially scrapped from the road network in 1985, Route 66 has long been a symbol of the American dream of freedom and roadside culture. The highway was built to connect urban and rural America, from Chicago to LA. The road to a better life cut across eight states and three time zones, paving the way for Thelma & Louise and Wyatt & Billy.

We too drive into the abyss in tandem. The Mojave Desert and the Painted Desert in Arizona are hellish, with temperatures reaching 45 °C in the shade, which is nowhere to be found. Steam is coming off both us and the bikes, which are red hot against our boots. And so we leave before dawn to get ahead of the heat. Now and then we whizz along dirt roads and single-track paths as if possessed by the dust, searching

Feeling small, flanked by redwoods on the Avenue of the Giants.

We ride through a dream that has been partly swept away. The road became a victim of its own success.

for a slither of shade and Route 66. Like so many who have gone before us. Because the 'Road to Opportunity' has been ridden with the courage of despair for decades. By Dust Bowl migrants in the 30s, Second World War soldiers in the 40s, and in the 50s and 60s thousands of Americans searching for freedom and attractions such as the Grand Canyon and Disneyland.

Today, that glory has largely faded. We ride through a dream that has been partly swept away. The road became a victim of its own success. In 1985, the new Interstate 40 brought the car-loving American from A to B even faster and more efficiently. Since then, all official road signs have been removed and each state decides for itself what it wants to do with the old road. So you can't really call the route well

signposted. Here and there the road splits into various side roads or loosely overlaps with Interstate 40. But for us, exploring all its nooks and crannies is part of the charm of the journey.

The many flourishing businesses that have sprung up along the route over the years have been trying to survive ever since, often in vain. What remains is a journey back in time past abandoned motels and gas stations, desert, mountains, rattlesnakes: Americana in all its glory. Road signs still lure you to take a pit stop at one of the many attractions, from dinosaur statues to trading posts to the

Sun Studio, Memphis. The birthplace of rock 'n' roll. *(left)* **Venice Beach.** *(top)* **Endpoint of Route 66, Santa Monica.** *(bottom)*

Tribute to Robert Waldmire
The Last Stop Shop The End 2448 Miles
MOTHER ROAD
SANTA MONICA
CALIFORNIA
ROUTE
66
THE END OF
2448 MILES
ROUTE 66
HISTORIC
CALIFORNIA
US
66
SANTA MONICA PIER
Official Headquarters
US
66
America's Main Street
The Mother Road

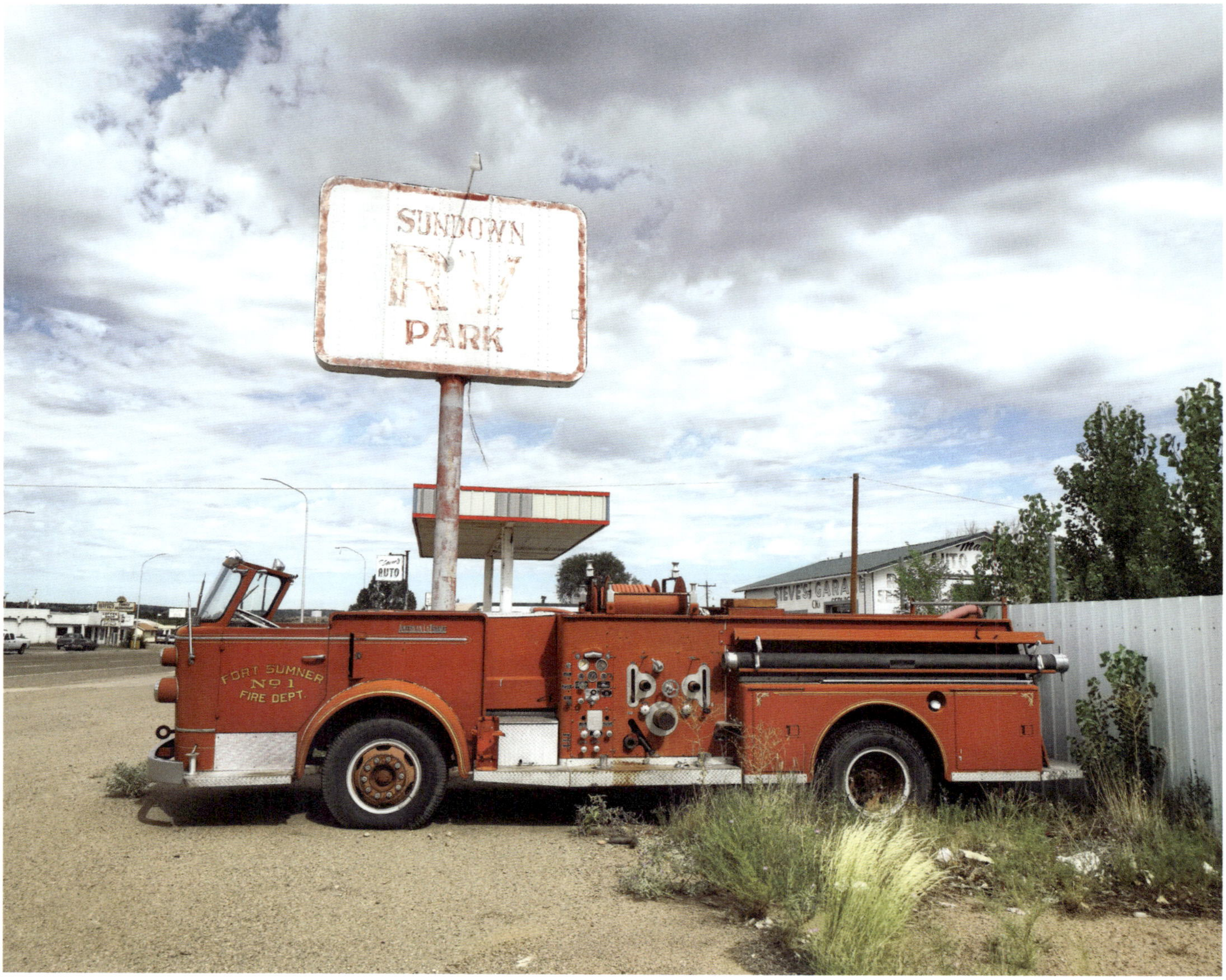
SUNDOWN
RV
PARK
FORT SUMNER
Nº 1
FIRE DEPT.

M AUW
628
Motocore

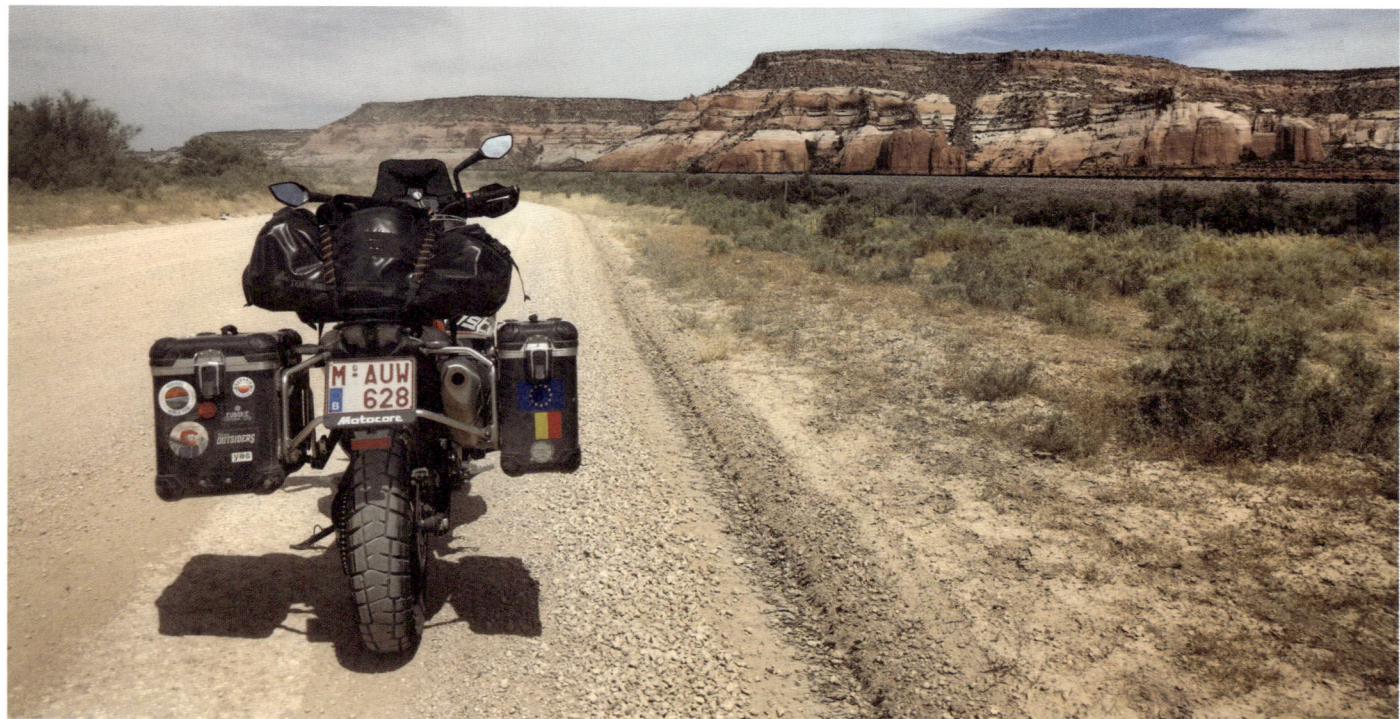

Sundown town references
remain along the Mother
Road. *(top)* Inland roads
through flyover country.
(bottom)

world's second largest rocking chair or best-preserved meteorite crater. They often seem like absurd artefacts from another era.

But not all that glitters is gold. Because the most romanticized road in the world also has a dark side, full of stigmas and stereotypes. Not everyone gets their kicks on Route 66. Much less known and talked about than milkshakes and rockabilly are the stories of minorities who also lived and travelled on this road. For example, African-American soldiers and travellers who weren't allowed to eat, sleep or refuel along this 'freedom road' for decades. For them, every mile became a minefield. They travelled with an emergency plan, cover story or driver's hat for disguise. A wrong stop at a public toilet, restaurant or hotel could lead to humiliation, discrimination or violence; an empty gas tank to a life-threatening encounter with the Ku Klux Klan. Companies with three Ks in their name were not a coincidence, but a code.

In an era of sundown towns, segregation and lynchings, the *Green Book* became an indispensable resource for those wishing to safely navigate the states. This annual guidebook with maps was published from 1936 to 1966 by Victor Green, a postman from Harlem. Also known as *The Negro Motorist Green Book* or the *Overground Railroad*, the *Green Book* listed hotels, restaurants, beauty salons, nightclubs, bars and gas stations where travellers of colour were welcome. In addition to safety, it also created economic opportunities for black-owned companies.

Along the road, more than half of which used to cut through the land of 25 tribal nations, more and more caricatured representations of Native American culture emerged. Anyone who really wants to honour Route 66 is entitled to the full spectrum of stories. Through initiatives like the 'American Indians and Route 66', 'The Women of the Mother Road' and 'Hispanic Legacies of Route 66' from the National Park Service, the road may just help dispel those stereotypes today.

Where do we go from here?

In Arkansas, we leave the 'Road to Opportunity' behind and cross the state towards Memphis, the starting point of the iconic Blues Trail. The Mississippi Delta is the cradle of all things blues, rock 'n' roll and soul. Presley's Graceland, B.B. King's and Johnny Cash's Sun Studio, a live jam session on Beale Street…

But Memphis isn't just the resting place of the king of rock 'n' roll. On 4 April 1968, Martin Luther King, civil rights activist and spokesman for the Black Freedom Movement, was murdered on the balcony of the Lorraine Motel. King was in town for the Poor People's Campaign and Salvation Strikers, a key march in the fight for equal rights. A museum was built around the motel, commemorating King and the crucial role he played in the Civil Rights Movement. At the National Civil Rights Museum, we get an overview of the history of American civil rights and African-American culture. A Route to Opportunity, after all. //

'I wanted to share the real story of Route 66–its broken promise of freedom. Black Americans who hit the road with a copy of the Green Book, *a guide expressly created to keep them safe in a wildly perilous landscape, surely already understood that the hopeful Mark Twain quote gracing almost every* Green Book *cover–'Travel is fatal to prejudice'–was purely aspirational.'*

Candacy Taylor, author of *Overground Railroad: The Green Book and the Roots of Black Travel in America*

View from Elvis's regular booth at the Arcade restaurant, the oldest in Memphis.

Buck Atom, 6-metre (19-foot)
tall space cowboy from Tulsa,
Oklahoma, Route 66.

LOCATION

Flyoverland
Central states, USA

CURIOSITY

Route 66 is the quintessential American road trip. Like an artery, the Mother Road nurtured communities and serviced millions of truckers and road trippers for decades. But not everyone got his kicks on Route 66... Uncover the many faces of the flyover states.

ADVENTURE TIME

Driving the 3,940 kilometres (2,278 miles) of Route 66 usually takes about two weeks when you include stops and visit the cities that are on the way.

TRAVEL TIP

The weather is fair during April, May, September and October, and the crowds are less abundant.

BEFORE YOU GO

Read into the myths and mysteries of the route, its strange roadside attractions, kitsch and living history.

No man's land

Stranded with engine trouble in Magdagachi

In my imagination, Siberia has long been synonymous with isolation and Arctic cold. The kind of place where adventures abound. With hardened inhabitants, all wrapped up in bearskins against the biting wind. A place you travel through, not one you stick around in for too long. Let alone one you want to end up stranded in. But in this despair, real life reveals itself. When you end up in a tinged reality with no way out.

The crescendoing chirp of crickets and birds has been announcing our arrival in Russia for miles. Our motorbikes wind their way around the peatlands that so characterize this sub-Arctic landscape. Flowering heather, mosses and birch forests come into view. The contrast with the barren plains and desert of the past few days couldn't be any bigger.

It's still swelteringly hot, though. Unlike its harsh winters, Siberia has hot summers for which it is less well known. Yet with temperatures fluctuating between minus 67 °C and plus 38 °C, it is also the official Guinness world record holder in extremes. And that temperature range is getting bigger, evidenced by the melting permafrost. A first misconception is shattered.

Our tyres have now covered more than 10,000 kilometres (6,214 miles) of European-Russian tarmac and Mongolian gravel, and we're just a stone's throw from Ulan-Ude, the capital of the Russian republic of Buryatia. From there we still need to wiggle the 3,000 kilometres (1,860 miles) alongside the Trans-Siberian Railway to reach Vladivostok. An iconic ride with more potholes than hot spots.

Lenin shows us the way east from Magdagachi.

The Far East

Our first night across the border transports us straight to the Siberia I imagined. Left and right, small wooden houses appear out of the dust and fog. Something wolf-like chases after us playfully. Children wave, their eyes widening at the sight of our bikes. Their mothers and fathers are more reserved. Nobody speaks English. The man who points us in the right direction is wearing little more than a dusty wife beater and the scent of home-brewed alcohol. In our room everything seems new, but the tap spits out brown water and the toilet has algae growing in it. A glitch in the matrix.

In the days that follow, we lose all sense of time and distance. Between the tiny villages of this Far East, there's a whole lot of nothing. When we ask in a little shop where we can get a cup of coffee, the woman behind the counter refers us to a café 50 kilometres (30 miles) away. An ATM? That's 60 kilometres (37 miles). For petrol, you'll need to hold on for 300 kilometres (186 miles), if you're lucky. We sleep in wooden huts, between the Siberian truckers. The tent isn't an option, because the ground is too soggy and the mosquitoes have taken over. There's often no running water, and no toilet except for a hole in the ground. The emptiness of Eurasia stretches your mind to its limits.

But we have more than just the sun to help us navigate. Everywhere we go, Lenin points us in the direction of Vladivostok. In village squares, in a cafeteria... even deep in the Siberian forests, he keeps popping up. More than 100 years after the end of the October Revolution, when the world's most famous revolutionary and first leader of the Soviet Union came to power, there are still lots of statues of him in Russia.

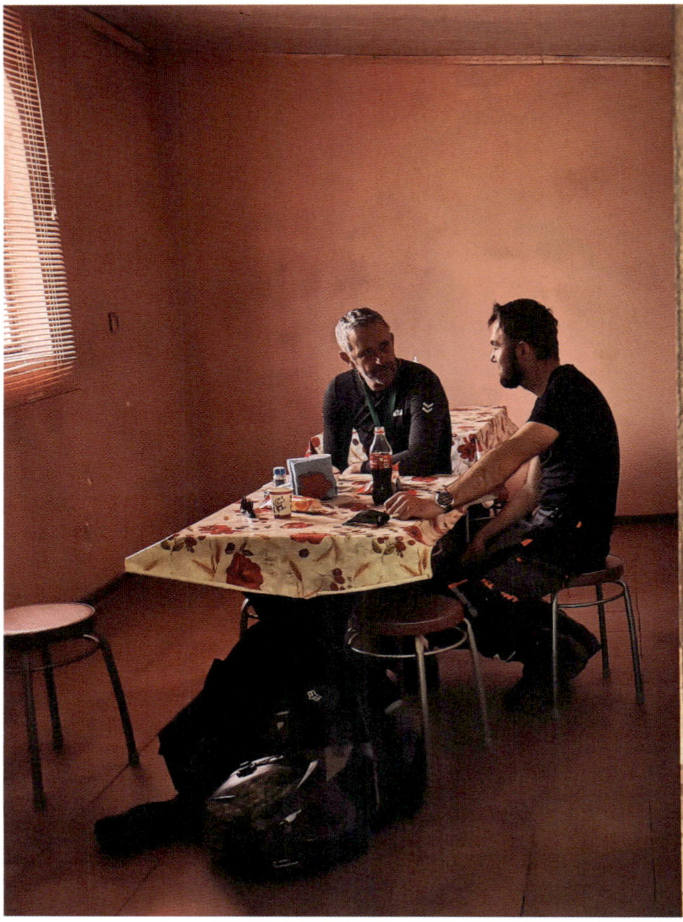

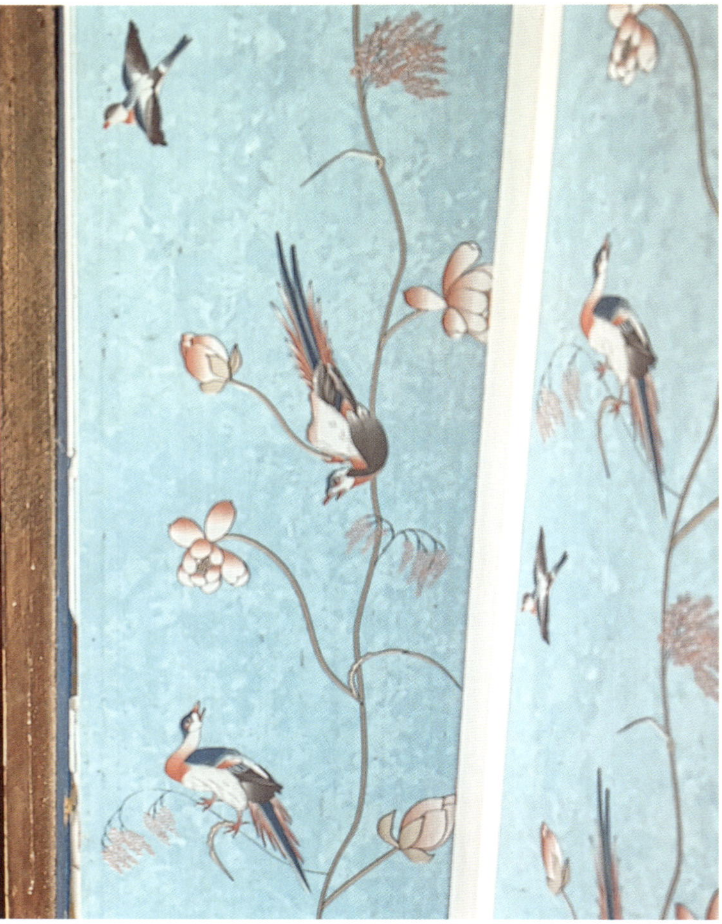

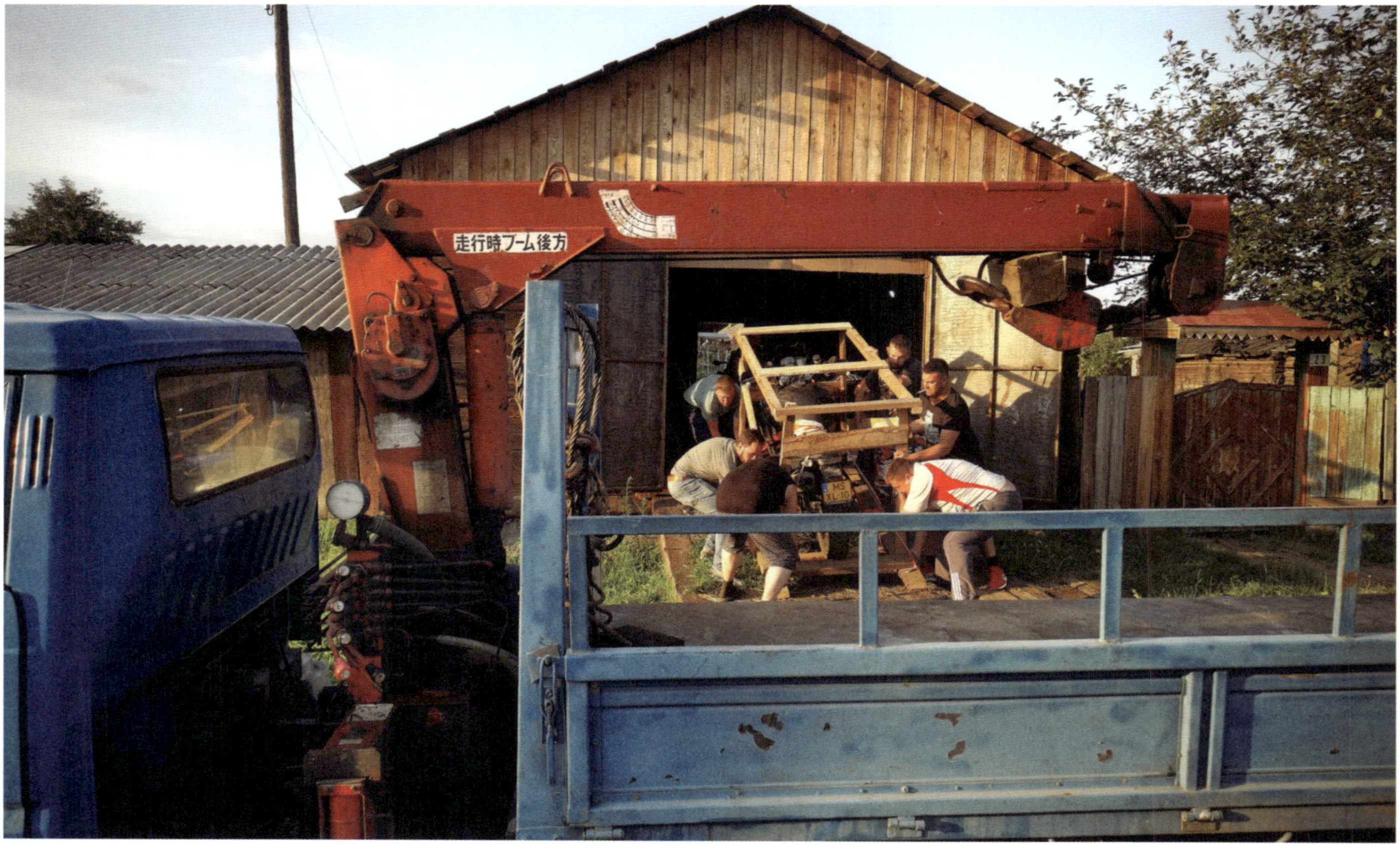
走行時ブーム後方

The Russians themselves don't really know what to do with them all. An increasing section of the population is speaking out more and more against their removal. And so the monuments continue to be maintained.

On the road, I keep myself focused by trying to decipher Cyrillic signs before they appear in my rear-view mirror. Even villages with just a dozen inhabitants welcome you here with a towering concrete structure. You'll also come across heroic monuments in all shapes and sizes. Warriors, cosmonauts, tanks, cannons, fighter planes, workers with clenched fists... The old Russians know how to make an impression.

Breakdown

We had just ridden through the same storm twice, when the unthinkable happens. The headlight, generator and battery of my travel companion's BMW give up almost simultaneously. As the storm breaks again, we break our cardinal rule: always stay together. Soaked, I stand guard next to the bike, while Ab goes in search of help. After an hour and a half of radio silence, he reappears in a small pick-up truck, itself punctured en route. The bike is tipped on its side and into the back of the pick-up. When we reach the local garage, by some miracle just 30km (19 miles) away, it's hoisted out again with a pulley. Nobody speaks English here either. There are no European spare parts and even having them sent out straight away would put us back several weeks. And so we end up stranded in Magdagachi, more than 1,500km (932miles) from Vladivostok. The kind of village outsiders would warn us about, and for that reason one we'd never have dreamt of stopping in.

The same Magdagachi where we're immediately offered a bed, a feast and a talent show. Where half the village springs into action to get us 'home' again. Because, after some painful dissection, it turns out the bike can't be repaired. And nobody can give us a ride. There's only one road for miles around, with no through traffic. We're driven around the village, without knowing where we're going. Plans are drawn in the air and in the sand. All around us, the villagers talk, discuss, deliberate

Russian roadstops. *(top)*
The crate of redemption.
(bottom)

and weigh up the options. We let down our guard for a while, and in blind faith leave our belongings and bikes behind with the keys still in the ignition. Thousands of miles from your own network, you have to take a gamble on the local heroes.

Convoy

In the end we were able to leave Magdagachi about six days later, on one motorcycle. A group of bikers had been rounded up to collect planks of wood and build a makeshift wooden crate for the other one. To the roar of a dozen motorbikes, we're led out of the village before dawn. We jostle about on the one bike behind the van they're using to personally transport the crate to Blagoveshchensk, a town on the Chinese border, 550km (341 miles) away. A point of honour. No arguments.

Yarzai – lawyer, biker and Zen Buddhist – is appointed as our personal bodyguard. He makes sure that the crate with the motorbike ends up on the right truck to Vladivostok, and talks non-stop about his city, an intriguing place with its Russian-Chinese mix.

In the evening, the Chinese city of Heihe beckons with a light show over the Amur. In winter, buses drive across the frozen river. Today, a ferry sets sail. Both Blagoveshchensk and Heihe were for a long time known for their so-called 'suitcase trade', where Russians would return with suitcases full of cheap Chinese stuff, which they then sold at a profit in Russia. For a long time, Heihe did everything it could to stimulate this form of trade, because it was its only raison d'être. The Chinese salesmen learnt Russian, translated signs into Cyrillic script, and even adorned their stands with faux-Soviet items so that the Russians would feel right at home. Today it's a tourist attraction for the Chinese who prefer to visit Russia in their own country.

The next day Yarzai leads the way for 350km (217 miles) to the border of his Amur Oblast. Another week later we reach Vladivostok, our most eastern point overland. Differently than planned, but even more fulfilling. Despite all the warnings, we got to know a warm-hearted interior. Never paint two Russians with the same brush. And don't judge a biker by their jacket. Even if it's adorned with Siberian tigers baring gold teeth. //

Riverside view of the Chinese city of Heihe, from Blagoveshchensk. *(left)* **Lada country.** *(following pages)*

LOCATION

Magdagachi
Siberia, Asian Russia

CURIOSITY

Siberia is home to the world's longest railway line and the world's deepest lake.

ADVENTURE TIME

It is unbearably cold in winter and boiling in summer, so the best time to visit Siberia is in March when the temperature varies from -5 °C to -25 °C. But don't let that stop you.

TRAVEL TIP

Visit Pleistocene Park, a 144km² (56 square miles) fenced-off area in Arctic Siberia, where Russian scientists are trying to stave off catastrophic climate change by resurrecting an Ice Age biome complete with lab-grown woolly mammoths.

BEFORE YOU GO

Few Russians speak English. Learn some basic Russian phrases and words, or decide to stick to interpretative dance and hand signals. Пожалуйста!

ON THE
WILD SIDE

Gorillas in the Mist

On the trail of the giants of the jungle

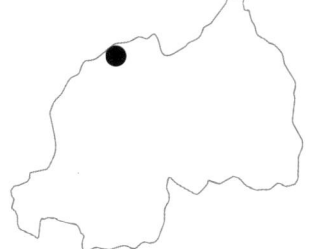

Welcome to the land of a thousand hills, and the home of the silverback gorilla. One of our closest relatives, it is one of the most endangered animals on earth.

Rwanda might not ring any bells as a travel destination, but this small, landlocked country offers a front-row wilderness experience. Our expedition is set in the foothills of one of many extinct volcanoes in the Northern Province. And it's not by chance that we've ended up at these coordinates: this is the exact section of the African jungle where the pioneering primatologist Dian Fossey undertook her extensive study of silverback mountain gorilla groups. She observed them in these forests over a period of 18 years, before being murdered in her jungle hut by an intruder (the case is still unsolved). And thus this is an iconic place, with a story as controversial as it is inspiring.

Under pressure

'Had Fossey not so fiercely protected the gorillas and their habitat, these apes, resting on the high-elevation slopes of Mount Karisimbi, probably wouldn't exist today', notes science journalist Elizabeth Royte. When Fossey began to study them in 1967, gorillas were perceived as savage and violent. She took it upon herself to change that image, and became both their spokesperson and guard for more than a decade. This fierce dedication made her a role model for aspiring conservationists. But her battling methods – including kidnapping and torture – often elusively described as 'active conservation', earned her the enmity of many.

Three decades after the groundbreaking researcher was killed, the ape population has grown from below 275 to a total global wild gorilla population of over 1,000 individuals. Ensuing conservation efforts from the Dian Fossey Gorilla Fund and the WWF International Gorilla Conservation Programme have found success. But the mighty mountain family is still very vulnerable, and ecotourism seems to have become both a blessing and a curse. It had such an accelerating impact on the growth of the species and the economy of the country, that it could become a victim of its own success.

I can't wait to see them myself, the wild animals we share around 98% of our DNA with. I must have browsed that boxful of pictures of Nyiramacibiri, 'the woman who lives alone on the mountain' – the locals' nickname for Fossey – a hundred times before I set foot in Rwanda. Being that fearless in the wild on your own, surrounded by animals, felt like living a childhood dream.

But we'll have to take it one step at a time from here. I join an armed team of trackers and anti-poachers who set off for an uphill hike on the slopes of the Virunga volcanoes. During the Rwandan genocide in 1994, rebels crisscrossed and randsacked the forest. These brutal events were followed by years of civil unrest. And since poachers are still active in the park, both rangers and tourists are vulnerable targets for kidnapping, or worse.

Jungle hike

Today, as in Fossey's day, the paths have to be cut through 1.8-metre (6-foot) tall grass, bamboo and eucalyptus forest. Gorillas prefer the path less travelled. After crossing the

Face to face with the great primates, the hair on my arms was standing on end.

As in Fossey's day, the paths have to be cut through 1.8-metre (6-foot) tall grass, bamboo and eucalyptus forest.

3,000m line (almost 9,900 feet), I try to enjoy the surrounding nature while panting for breath. The higher we get, the more sounds become our guide. We hear the animals around us well before we spot them. And vice versa. The dense underground and steep, slippery trail soon has us scratched, muddy and exhausted. I feel humbled already.

On the way up, our guide Oliver introduces us to the secrets of gorilla language. We've got clear instructions on what to do and what to avoid. The silverback mountain gorilla is high-listed on IUCN's Red List of Endangered Species, having survived poaching, nearby wars, disease and habitat loss. We don't want to stress them out, or feed or touch them, because they have little immunity to human diseases. And, I bet you wouldn't want to bump into one too, given the fact that silverbacks are even larger than lowland gorillas, weigh over 130 kg (300 pounds) and have imposing big faces.

After a few hours of hiking, our guide urges us to be prepared and on our guard. For our untrained eyes, it's almost impossible to spot the primates in the thick vegetation of the dense rainforest. Luckily, our guide and his trackers notice every knuckle print, bent stem and absent leaf. One of them is a former gorilla poacher who changed sides. He has been following different gorilla families from dawn to dusk for a few years. From here on, every sound makes my heart pound faster.

Furry encounters

We keep climbing onto the ridge-line that joins the Sabyinyo and Bisoke volcanoes. It takes us about five hours to find our first gorilla's nest, in a lofty spot in an ancient volcano. And then, I'm stunned by the scene's beauty. When the first black fur appears through the

The road to Mount Karisimbi. *(left)* **Face to face with the imposing silverback Munyinya.** *(right)*

jungle greenery, it's magic. A chubby belly here, an arm and a couple of baby toes there. We hold our breaths collectively. There's no sound but the wind, some soft gorilla snores and an occasional grunt. The Hirwa family consists of 12 individuals. One dominant silverback and its son, five adult females, two sub adult females and three babies. They are named for the way the black hair on their backs changes to silver as they age. When one of them turns around and rises to beat his chest, the first eye contact makes me freeze. What an intimidating beast. In the flesh. Only a short distance away. Yikes.

Hirwa means 'the lucky ones' in the local language. Wondering who are the truly lucky ones among us, we observe these free-roaming great apes for more than an hour. We stay put and imitate some low nasal sounds, until we are no longer

Like human fingerprints, these nose wrinkles are unique for each gorilla. Researchers use them to identify individuals while observing them in the wild.

of interest to them. Fossey was one of the first to develop the technique for gorilla habituation, by mimicking the social cues of the apes to gain their acceptance.

The head of the clan is Munyinya, an imposing 180 kg (400-pound) silverback who wandered through Congo and Rwanda before he settled here with a few females. You can identify each individual by its distinctive nose print. One of the rangers collects some poo samples for the lab. While the mother gorilla crushes some ferns, the little ones get up to all sorts of mischief, including the cutest all-mighty chest beats ever. All in all, they are just like us… When our time in their habitat is up, I leave in amazement. Feeling a little guilty for invading their privacy. This is a wildlife experience as pure as it gets. //

Curious adolescent gorilla. *(previous pages)* Eucalyptus leaves finding their way to the village. *(right)* Family outing. *(following pages)*

LOCATION

Volcanoes National Park
Rwanda, Africa

CURIOSITY

Researchers use nose prints of gorillas to identify individuals while observing them in the wild. Like human fingerprints, these wrinkles are unique for each gorilla.

ADVENTURE TIME

From mid-May to mid-October, the long dry season has perfect conditions for tracking gorillas.

TRAVEL TIP

Visit the Dian Fossey Gorilla Foundation to get an in-depth perspective on Dian Fossey's 18-year-long gorilla study as you trek through her living laboratory in the park.

BEFORE YOU GO

Take a deep dive and watch Dian Fossey: Secrets In The Mist, a three-hour series for National Geographic Channel. The series tells the story of her life, work, murder and legacy.

Puffin Paradise

Discovering Mykines Island

Misty mountain slopes, green grass and small villages with just a few inhabitants. The world's oldest archipelago is a place without self-importance, traffic lights or souvenir shops. The Faroe Islands are a safe haven in the middle of the North Atlantic Ocean, where front doors are rarely locked, and where there are more sheep than people.

The ship's horn sounds. The *Norröna* ferry is taking us from Hirtshals in northern Denmark even further north. It is the only ferry that stops in the Faroe Islands, on its way to Seyðisfjörður in east Iceland. A scenic 36-hour trip on the North Atlantic that gets me into the swing of things. Lying on the bunk bed in the cabin, or with a book on my lap and a Føroya Bjór beer in my hand on the upper deck. I've just finished reading Johan Harstad's *Buzz Aldrin, What Happened To You in All The Confusion?*, and I can already see the book coming to life. The main character also spent hours floating about on this ocean.

There is a total of 18 Faroe Islands at the centre of an isosceles triangle between Norway, Scotland and Iceland. This 55-million-year-old Jurassic World has a varying temperament and two faces. To the west, the landscape provides upright resistance against the storms and waves. To the east, islands with deep fjords beckon to the European mainland.

Gjógv, the village where the main character from Harstad's *Buzz Aldrin, What Happened to you in all the Confusion?* found his wooden sheep, shed and cliffhangers.

Greys & greens

A day and night later we sail into Tórshavn, the capital of the Faroe Islands, named after the Norse god Thor. A happy arrival, catching a glimpse of this colourful city and the first green fjords. I drive the car out of the belly of the boat, straight into town. Over the next few weeks we'll be hopping from the eastern to the western islands on small ferries, through narrow tunnels with passing spots for oncoming traffic. The locals look a bit surly at first, but are warm-hearted and very resourceful. Officially they are still part of Denmark, but they have their own parliament, language and free will. After Google refused to provide them with street view, they enforced their own Google Sheep View using dash cams on canoes, boats and thousands of sheep. The kind of quirkiness I'm immediately attracted to.

It's no coincidence that the name 'Faroe Islands' means 'Sheep Islands' (in Danish, *får* means sheep and *øer* means islands). Roaming freely from east to west, there are a total of 80,000 of the woolly creatures. That's 20,000 more than the total inhabitants in the entire island group. As well as green roofs for the sheep to graze on, you'll also find trolls (in Trøllanes) and a selkie (in Mikladalur). Just like the Icelanders, these islanders have plenty of great stories to tell.

Seeing red

We spend our first days exploring the islands from east to west. By car, boat and on foot, come rain or shine. Under the radiant sun, we explore the black beaches around the church of Saksun during a long trip along old fisherman and postman trails. In Klaksvik and on the island of Vidoy, we can't see more than a metre (3 feet or so) in front of us for three days. The impressive Cape Enniberg is also shrouded in a thick fog. It only takes a bit of wind to go from splendour to disappointment.

Sea kayaking to the rhythm
of the Atlantic Ocean. *(top)*
Cabin life. *(bottom)*

And that volatility also applies to the water. The tidal currents between and against the islands are deceptively strong and the water here barely reaches 11 °C in summer. Nevertheless, on Streymoy I want to go out into the North Atlantic, whatever it takes. At the highest point, winds reach up to 11 knots, leading to dangerous seas and giant waves at the entrances to the fjords. But that doesn't stop us. We each put on a thick drysuit and woolly hat, warm up our muscles and pull the sea kayaks into the water.

Our trip starts in Hósvík, on the east coast of Streymoy. The bright red suits and kayaks contrast against the green of the cliffs and the blackened basalt rocks. The salty wind scrapes my cheeks as the noses of the kayaks make their way through the ocean. It feels like paradise. Until we paddle past Við Áir, one of the world's last three whaling stations still standing. Closed in the 80s, it is now part of a maritime museum. The area is also one of the locations of the controversial *grindadráp* (whale slaughter) drive hunting. Dating from the 9th century, this traditional practice (now heavily regulated) turns the Sundini sound red, with hundreds of pilot whales and dolphins rounded up and killed. A horrific practice, which I immediately condemn. Hunting was once a matter of life and death for families on the remote islands, but it still survives as a tradition today – despite significant protest from animal activists. Our guide also doesn't want to rock the boat. We take a break, shake hands again over a hot cup of tea, then finish the kayak trip in silence.

Atlantic puffins

In the end it is Mykines that appeals to my imagination the most. The steep cliffs provide a safe haven for millions of seabirds that live here permanently, come here to

breed, or are just passing through. Of the 305 species, the petrel and the puffin – or more precisely, the Atlantic puffin – are the most iconic. After a bumpy crossing and a trip from the harbour, we reach Mykineshólmur via a footbridge over a gorge that's 35 metres (115 feet) deep. I feel like both Indiana Jones and Melanie from Hitchcock's *The Birds*. The small lighthouse island is the breeding ground for one of the largest colonies of Atlantic puffins in the world. They look both comical and sad at the same time, with their brightly coloured beaks and fish in their mouths. Millions of puffins fly to and fro to dig their nests into the cliffs, hatch their eggs, and around 40 days later disappear again with their young into the northern sun. It's as if we're in a nature documentary. //

'I had a job that night. I filled a vacuum. I was green. A small green spot in the blue painting. I was the deserted island that makes the ocean look so big.'

From *Buzz Aldrin, What Happened to You in All the Confusion?*, by Johan Harstad

The green wilderness of Saksun. *(right)*

Land of fog, sheep and
green roofs.

LOCATION

Mykines Island
Faroe Islands, North Atlantic Ocean

CURIOSITY

Sheep (and puffin) easily outnumber humans in this treeless
place. This gives a unique feeling to the islands, with never-
ending rolling hills interspersed with sharp, choppy outcrops
and fairy-tale houses.

ADVENTURE TIME

During the summer months of May to August, the weather
– and surrounding ocean – is calmer and the Arctic puffins and
other migratory seabirds arrive on the islands to breed and nest.

TRAVEL TIP

Bring your tent, camp on the one and only campsite of the
island and wake up surrounded by hundreds of thousands of
breeding birds.

BEFORE YOU GO

Roam the islands from home, thanks to Sheep View 360. The
Faroe Islands has become the latest remote part of the world to
be featured on Google Street View – thanks to one woman and
five camera-mounted sheep.

Winding fishermen's roads.
(left) Mulafossur Waterfall
in Gasadalur. *(top right)*
Atlantic puffins. *(bottom right)*

Wild horses

With the shepherds in the At-Bashi mountains

18

As the traffic jams build up on Mount Everest, the north-western foothills of the Himalayas remain silent. Kyrgyzstan, Кыргызстан or the Kyrgyz Republic is a relatively young nation, and perhaps the least known country in the world. Very few people seem to know how to pronounce it, let alone where it is exactly – even *the New York Times* once mistakenly called it Kyrzbekistan. Here, you'll come across Lenin, yaks, yurts and mosques. The snow-capped peaks of the Tian Shan mountains dominate the landscape at more than 7,000 metres (23,000 feet), while its valleys were once a haven for hundreds of thousands of Silk Road travellers.

It is precisely this mystery that makes the land of Heavenly Mountains even more appealing. A higher dimension that you can only really enter on horseback, in the footsteps of a shepherd. A journey of over 100 kilometres (60 miles) through the thin mountain air. Anticlockwise into the nomadic lifestyle.

'Chu, chu!' yells my guide Talgat as he passes by, patting my horse on the buttocks. My nameless stallion snorts and responds. I sink deeper into the wooden sheepskin saddle. Together, we leave the village behind us and trot towards the first ridges. Beyond the snow-capped peaks there's a rumbling sound. Now and then a crack and flash of lightning. The horses remain steadfast but we're not so much at ease. The dark sky makes the spectacle

Woolly goods at the animal market in At-Bashi.

even more dramatic. Today we'll travel around 25 kilometres (15 miles) deeper into the mountains, where I'll be dropped off at a shepherd's family by documentary photographer Frederik Buyckx.

This was once the beating heart of the Silk Road, which connected East with West. A transit country for traders, pilgrims, Buddhists and warriors. A string of peaks and valleys, the size of the United Kingdom, tucked in between China, Kazakhstan, Uzbekistan and Tajikistan. More than 90% of it is mountainous. Almost half is higher than 3,300 metres (10,800 feet), three-quarters of which lie under eternal snow and glaciers. In short, an area that's slow to cross, due to the state of the road – or lack thereof – and the locals' attitude to time. Still heavily depending on nature's cycles, people here often live semi-nomadic lives, without electricity, phone reception or running water. And although I also spot solar panels, smartphones and the odd moped along the way, since the collapse of the Soviet Union in 1991 it is mainly their own traditions that have gained ground, with the horse leading the way. Under Soviet rule, the Kyrgyz horse was crossed with larger Russian breeds to produce meat for slaughter. Traditional horse racing and games were abolished, as were the travelling troubadours and their traditional music. The modern Kyrgyz are once again proud of their nomadic lifestyle, language and clan. Even when Soviet campaigns forced people to live and work in villages and collective farms in the 30s, many continued to migrate seasonally with their herds and yurts. To this day, sheep, cows

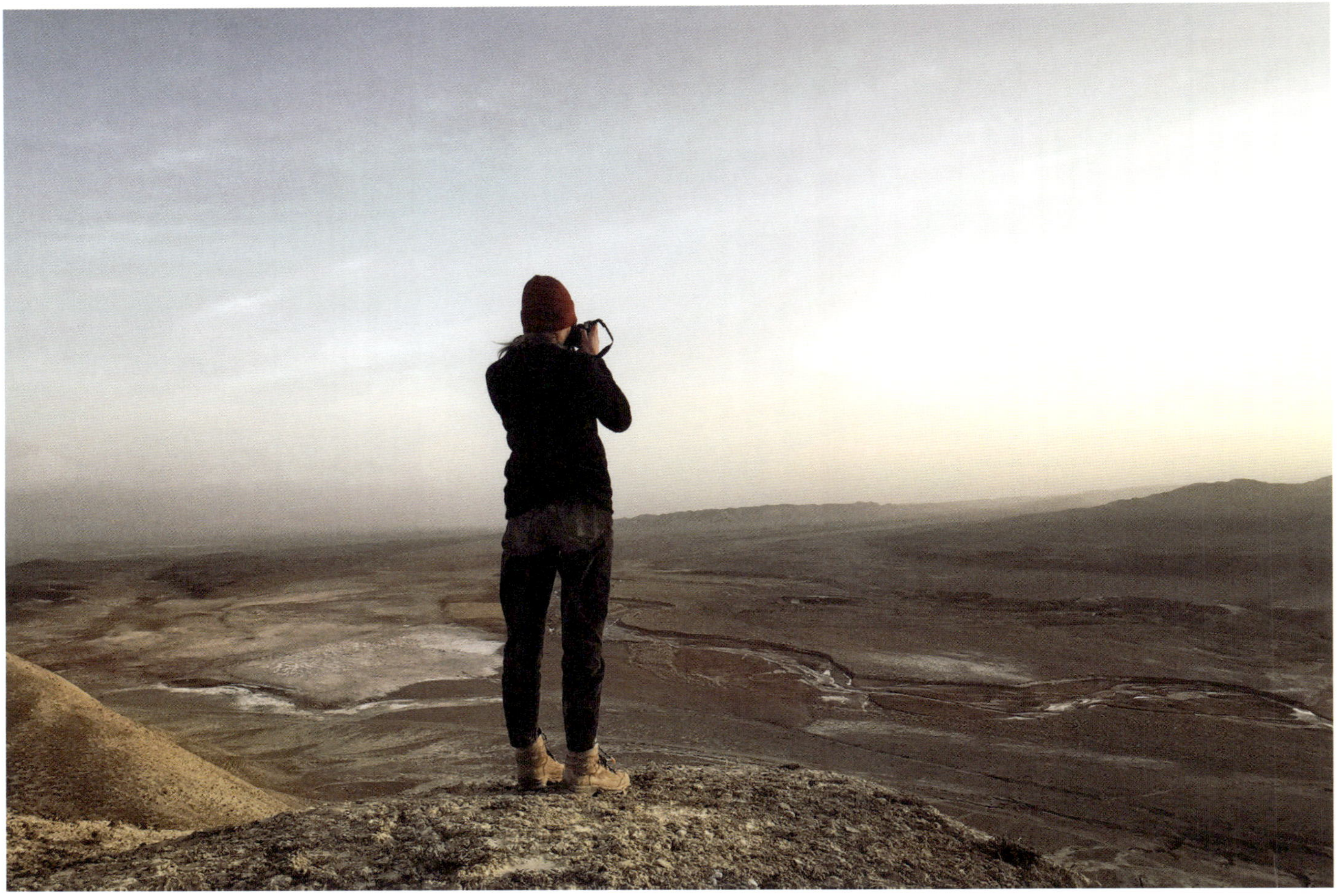

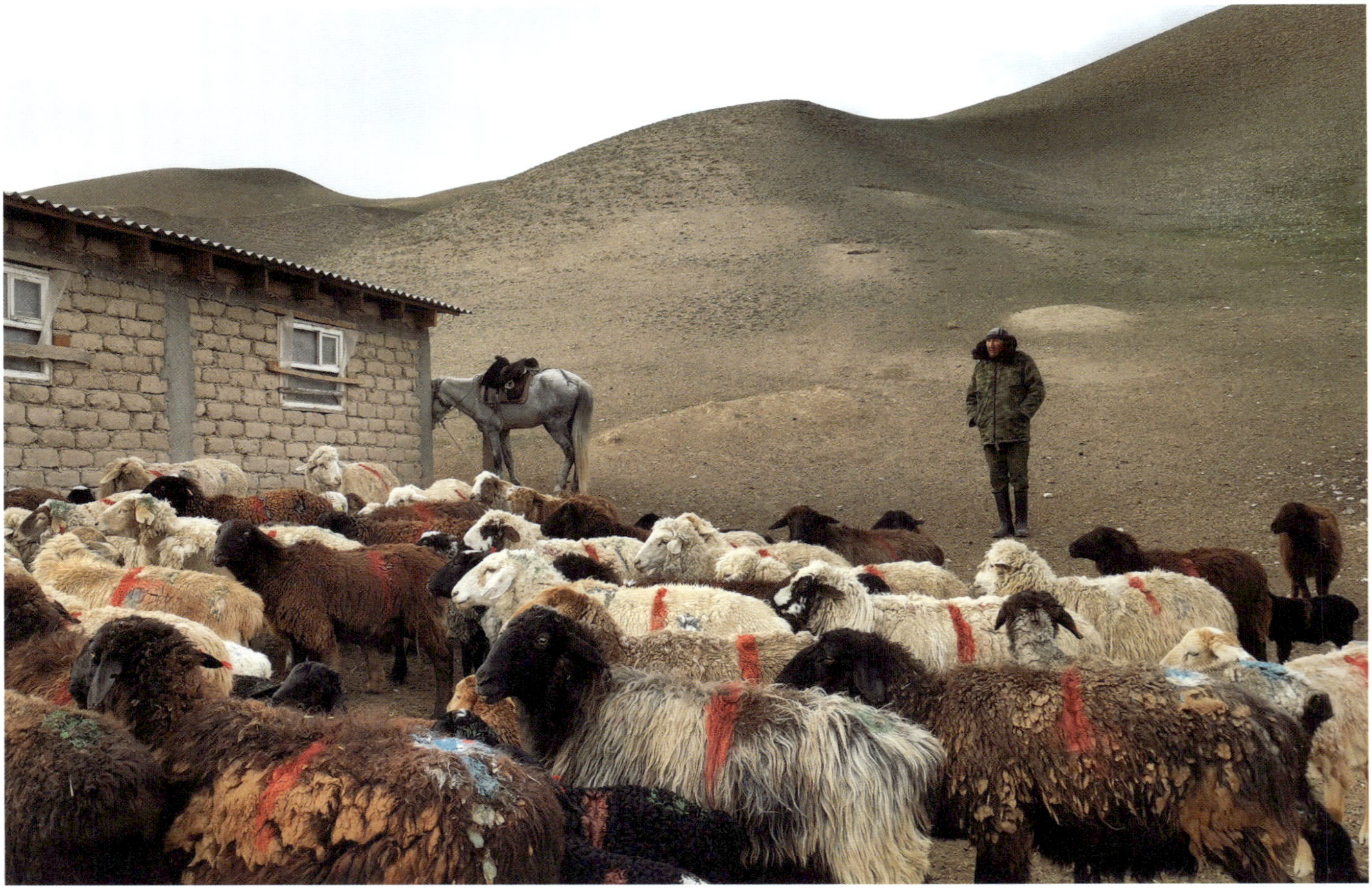

The best of Iceland, Switzerland, Mongolia and northern India unfolds before me, in flurries of velvet green, hints of orange, and zebra stripes.

and horses are still the most precious commodity. For most of their lives, animals here lead a fenceless, free existence, until they're roasted or ridden with respect.

Here, foals are born in the wild and children in the saddle, or so their fathers claim. Sadly, I don't share that natural talent. But fortunately, even on the steepest, most crumbling terrain, the horses prove to be miraculously steadfast. I undergo a first exercise in mindfulness on a perilous Looney Tunes slope. Too unsteady to climb, too steep for the horse to descend. Let's just zigzag then? We make our way around boulders, through mud and across river beds.

Mountain life

I'm dropped in a remote hamlet, welcomed by a man whose wide smile reveals all five of his teeth. My hosts, a shepherd called Avtandil and his family, only speak a few words of Russian, but we get by with some humming noises. I get a newborn lamb thrust into one hand, and a stick into the other, so that I can count the sheep coming down the mountain by the dozen. Job done, the family expresses their hospitality with naans, dumplings and litres of chai. We continue to explore our common ground under the light of a headlamp. Joining the three generations around the table, I turned out to be the long-awaited fourth player: for hours, we play the same card game, the rules of which will forever remain a mystery to me. As compensation I'm offered kymys, fermented mare's milk that's slightly alcoholic. Is it the altitude, the intoxicating smell of the sheep dung in the stove, or just tiredness? The interplay creates the kind of hilarity that can't be subdued.

Looking out over the valleys. *(top)* **Toktobek with the herd.** *(bottom)*

For a few days I take in the idyllic scenery, meandering

through the mountains on horseback with the shepherd and his herd. Out of bed early, galloping up the ridge, guiding around 200 sheep through the next valley. The language barrier means that we don't exchange many words. As with his dog, a nod or a whistle is enough. While herding I learn to pass the time and start seeing the lines in the landscape in a different light. I count cloud types, marmots and dozens of birds, from large birds of prey, such as snow vultures, kites and falcons, to strange apparitions like the crested hoopoe. The higher we go, the sparser the vegetation. At the top, all that remains is grass, wild thyme and miniature flowers. After that there is only the grit of the pale mountain itself. Here, a selfie at the tree line beats a footprint in the snow.

It's spring, but I can easily tick off all the seasons. The wind has picked up and the rain now lashes against my face. I don the heavy turquoise poncho, which looks more like a tent, tie the horse's front legs together so it doesn't stray and sit next to it while it grazes to its heart's content. An hour, maybe an hour and a half passes. The snow settles on any bit of space the mountain ridge has to offer, accentuating the texture of the rock. No flakes stick to the steep orange walls, making them extra unapproachable.

When the sky clears, I can see for tens of miles. The best of Iceland, Switzerland, Mongolia and northern India unfolds before me, in flurries of velvet green, hints of orange, and zebra stripes. During the Soviet period, this eastern part was off limits to outsiders, extra protected because it borders China. What a shame. Today the most beautiful mountains, their wolves, bears and lakes are once again accessible to all.

Dead goat polo

On Sunday we meet again at a more central settlement

I'm dropped in a remote hamlet, welcomed by a man whose wide smile reveals all five of his teeth.

in a valley, where shepherds claim our horses for a game of Kok Boru, a spectacle that's also called 'dead goat polo'. Before we know it, it's lost its head. And that turns out not to be a metaphor. A prayer is heard, followed by a quick beheading. Its lifeless torso is brought onto the pitch as the game's 40-kg (90-pound) ball. What at first seems like a barbaric frenzy turns out to be an age-old tradition among nomadic tribes, and one of Central Asia's most popular sports.

A rider spits at my feet and gallops onto the pitch. From the outside it looks like pure chaos, but everyone has their eyes on the prize. Two teams compete for the carcass three times for 20 minutes each time. The riders have to throw it into one of the round kazans as many times as possible,

while the horses continue to pound each other as if possessed. Clothes rip to shreds. Clouds of dust rise and disguise blood, sweat and falls. My jaw, front teeth and thumb throb with every shove. Solidarity, after a thud from the horse that's now stealing the show on the pitch.

There are different stories about the origin of the competition. Kok Boru literally means 'grey wolf'. Wolves that threatened the herds used to be hunted and killed by the men of the village. On the way back, they tried to steal the carcass from each other, because whoever could throw it into the elders' yurt when they arrived

Blood, sweat and tears during a game of Kok Buro. *(left)* Transported into the mountains. *(right)*

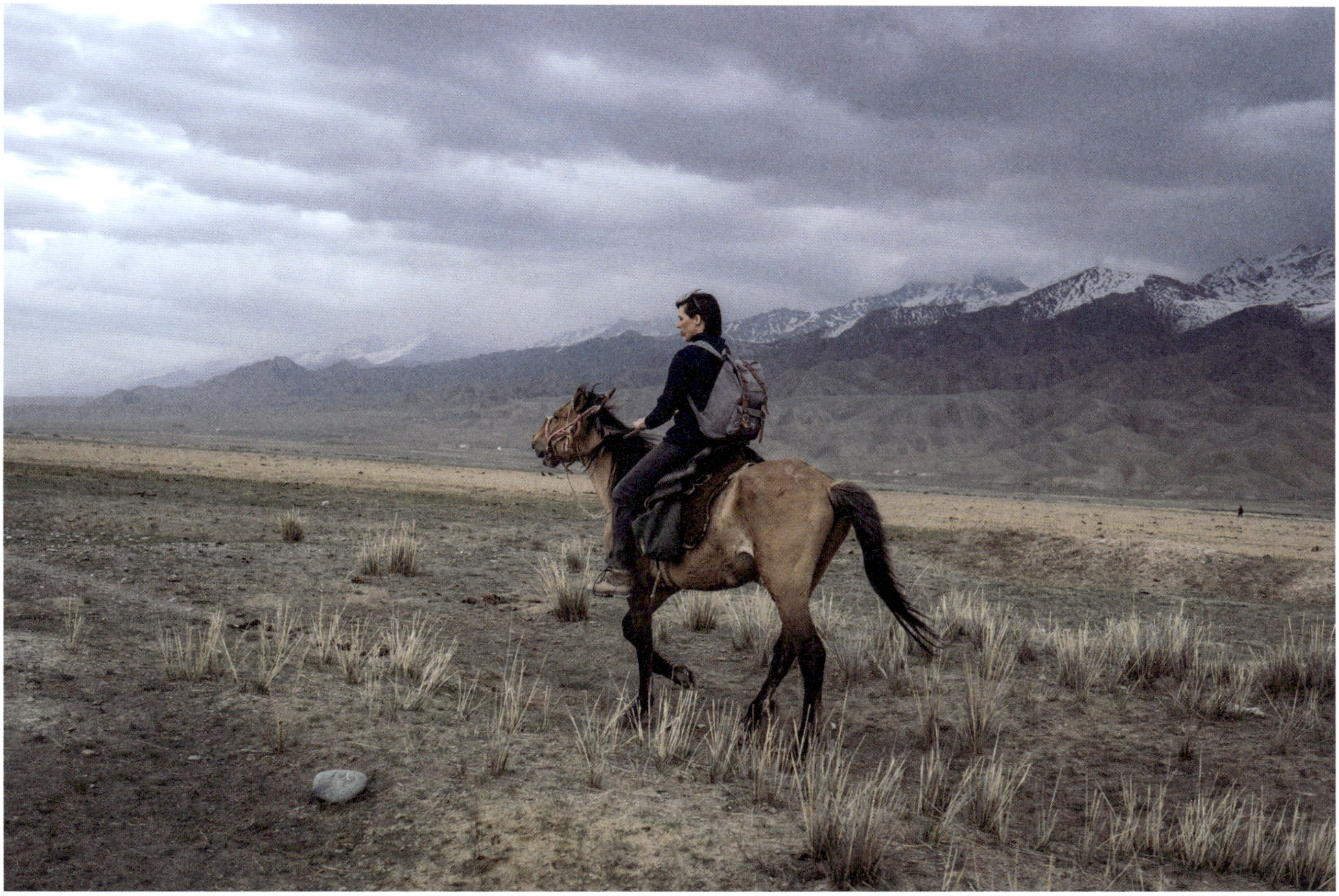

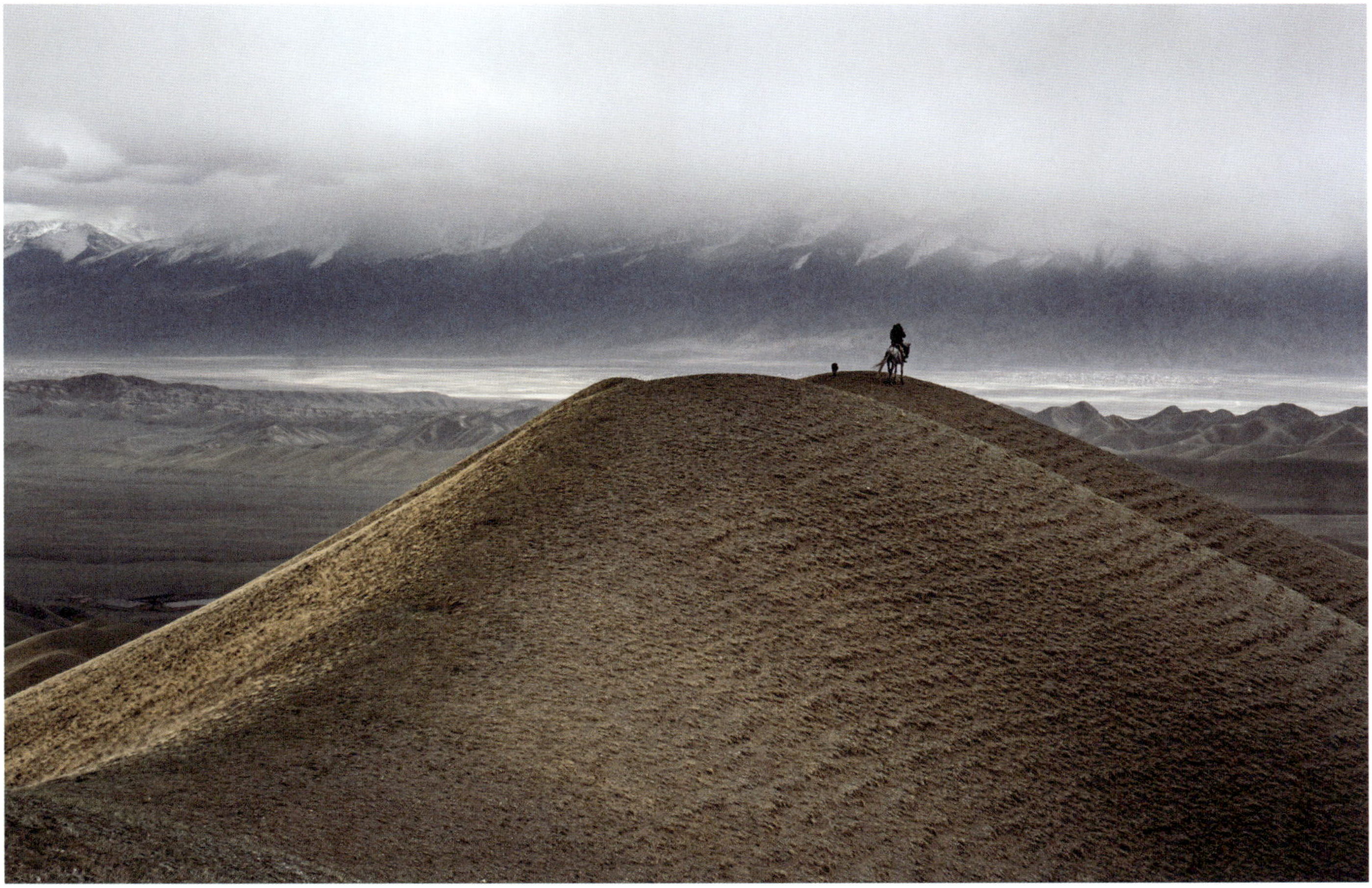

would win the respect of their clan. According to others, this is how riders and horses were prepared for relentless battle.

In the evening we have the honour of tasting the pummelled-flat goat meat, which has been simmering on the stove for hours since the end of the game. The contest and the feast loosen the tongues of the silent shepherds. A spontaneous boom-car party breaks out under the starry sky with singing and dancing in the headlights of a Lada Niva.

Village of death

Weeks of lonely altitude give you a God's-eye view of life. Before we drive back to civilization, we first have to pass an abandoned and overgrown village. The village of the dead. Ancestral cemeteries are deliberately built here on a hill or along the side of the road. 'Nomadic people want an overview and a view, even after death', says local farmer Aman. Fascinating not only for their beauty, but also as a blueprint of Kyrgyzstan's multicultural past. In the blink of an eye, I spot nomadic, Muslim and Soviet references alike. Their peaceful coexistence here is unique, as many of Kyrgyzstan's important mosques were destroyed during the Soviet era. An intriguing mix that can be felt from the capital's main square to deep in the mountains. //

Traditional graveyard in a mountain village near At-Bashi.

Whiling away the time with
a view of horse manes and
valleys.

LOCATION

At-Bashi mountains
Kyrgyzstan, Central Asia

CURIOSITY

The Kyrgyz people produced the longest poem in history. With a whopping 500,000 lines, this epic poem about the life of the warrior Manas is even 20 times longer than the Odyssey.

ADVENTURE TIME

The best time to visit Kyrgyzstan is from April to September, when it is generally dry with low amounts of rainfall. Best explored in 2-3 weeks.

TRAVEL TIP

There is nothing more natural than riding a horse in Kyrgyzstan. If you're up for a real adventure, join 'The Dusty Shepherd' for a horse trek in the mountains. Photographer Frederik Buyckx knows the vast and barren landscapes of the region inside out, and will allow an intimate glimpse into the shepherd's lifestyle.

BEFORE YOU GO

Read *Jamila* by Chingiz Aytmatov, a story set in Kyrgyzstan around the time of the Second World War, praised as 'the most beautiful love story in the world'.

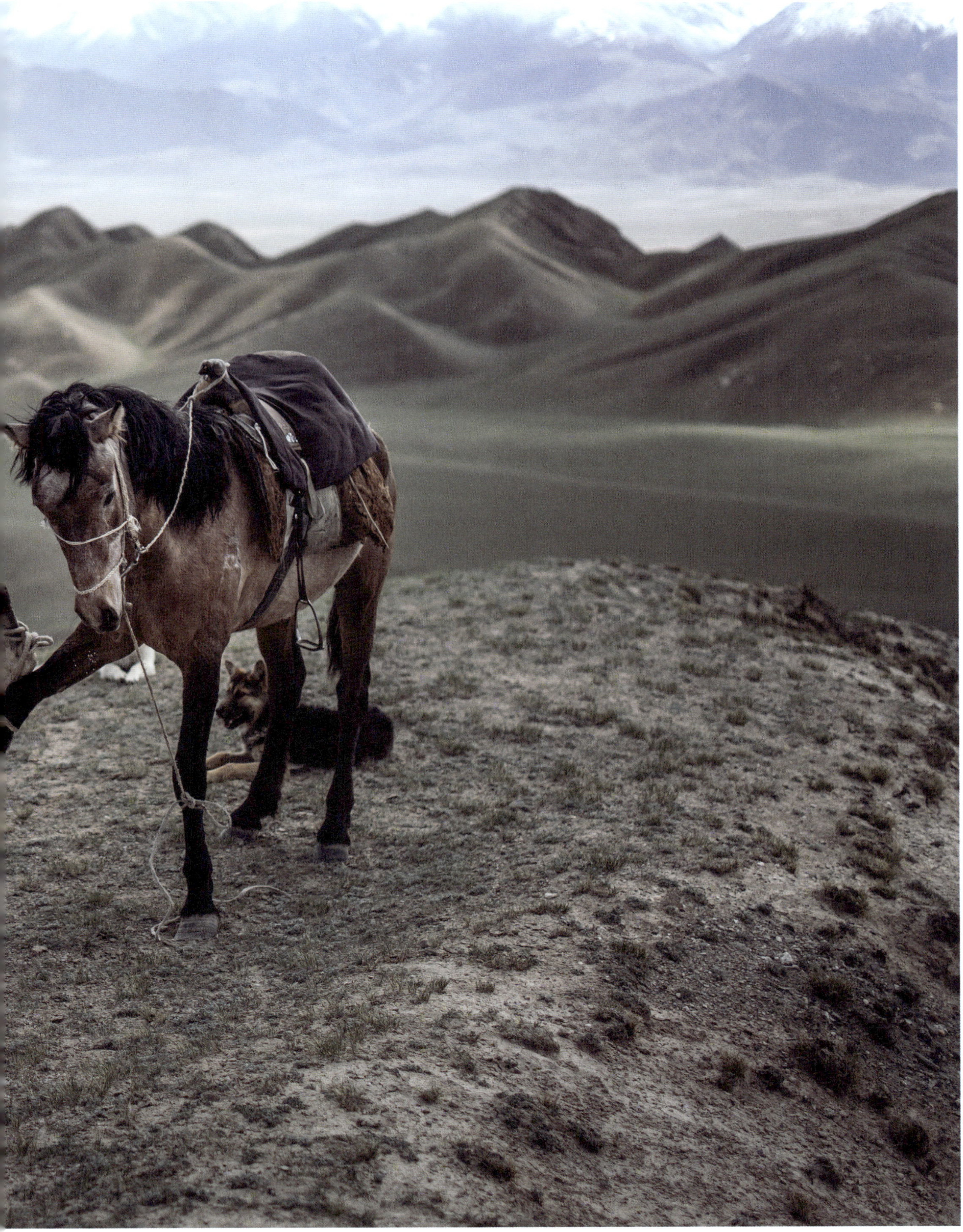

Rewilding a Nation

Rediscovering the primal force

19

Around crowd-pullers such as Loch Ness and the Isle of Skye, Europe's last wilderness is thriving. Or will we soon be able to say 'first'? Because the Scottish Rewilding Alliance is committed to making Scotland the world's first official Rewilding Nation. Dozens of like-minded landowners and organizations are working on restoring nature on a large scale. A wilderness in which animals can move freely. I'm also going to explore it, from south to north. A 1,290km (800-mile) hike taking in mountain peaks, beaches and ruins.

My road trip starts in the Cairngorms National Park, the wild heart of Scotland for which Thomas MacDonell, director of conservation at Wildlands, a Scottish conservation organization founded by Scotland's richest man, has devised a 200-year plan. A timeline that transcends his generation, and offers room for a drastic reinterpretation of the Scottish landscape. Because that landscape turns out to be less pristine than you might think: Scotland's now-familiar moors were once covered with primary forest, which quickly disappeared in the 19th century as trees were cut down to build merchant ships. And with those forests, a lot of flora and fauna also disappeared. So it's time for change, and a journey back in time.

Glen Affric

I open the door of the old Defender and hoist myself into the adventure. Quite literally – some people need a rope to pull themselves into these seats. From here we drive north, and in this workhorse we quickly blend into the

The Scottish Highlander in its equally impressive habitat.

landscape. The dashboard consists of two analogue dials and a series of universal, black buttons. Baking hot or freezing cold are the two temperature settings. We sleep wherever we end up, in a canvas rooftop tent. This gets off to a great start: just before sunset we find a secluded spot in the middle of breathtaking Glen Affric, where we set up the tent for the first time. I clamber up the car to make it wind- and waterproof, and while I'm up there I get an even better view of the valley. The fog lifts over the heathland. I hear a buzzard, a sparrowhawk and a bellowing deer. Around the riverbank, pines grow alongside golden birches, flanked by ferns, flowers and twisting plants. Everything is both magical and mildly terrifying, providing an evocative picture of what much more of the Scottish Highlands could look like.

This landscape is a sample of the catch-up movement that has only really just started: Scotland isn't yet ready with its test case. The country has been experimenting for several decades with large-scale rewilding projects that return huge estates to nature. Animals that had disappeared, such as the wolf, lynx and ungulates, are being reintroduced. Successfully so, because more than 5,000 animal species are already flourishing here. And we can attest to that. In the morning we spot a red deer bounding off, just after being woken up by some Highland cattle that came to scratch themselves on the ladder of the roof tent. MacDonnell calls the ungulates the sculptors of the forest.

Jurassic Park

We continue our way north, via the rugged west coast. Here, too, the rewilding has started. The single-track roads of the North Coast 500 lead to villages such as Kinlochbervie

In the morning we spot a red deer bounding off, just after being woken up by some Highland cattle that came to scratch themselves on the ladder of the roof tent.

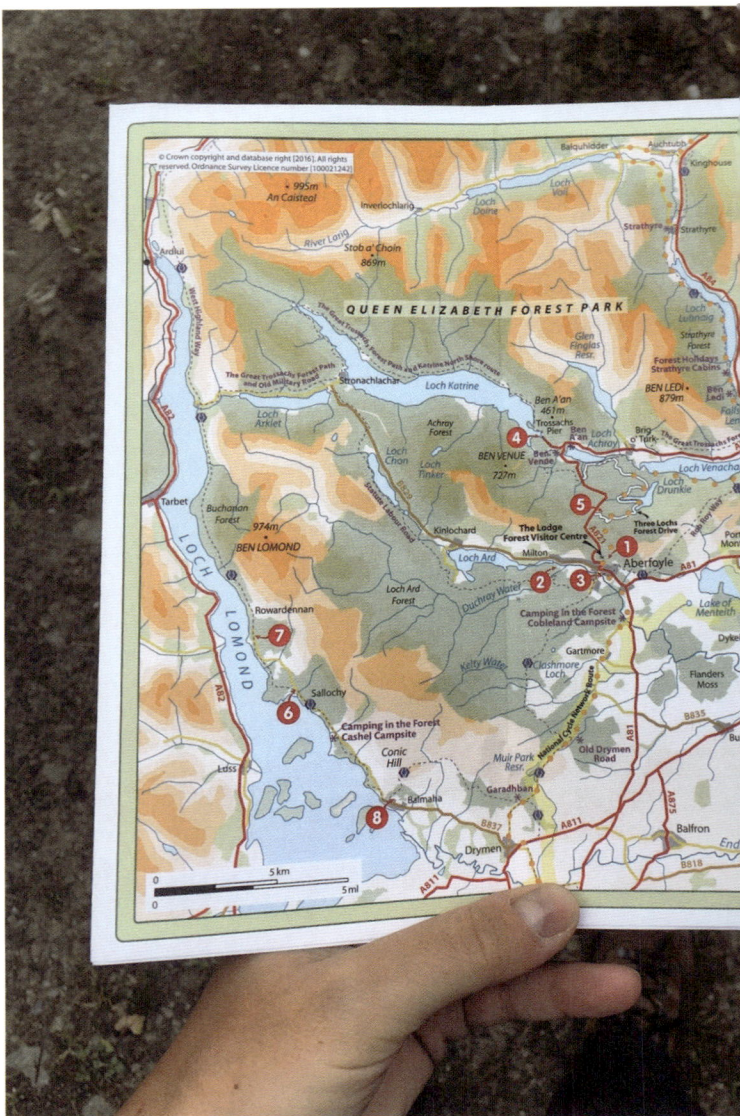

– Ceann Loch Biorbhaidh in Scottish Gaelic – which has more syllables than houses. There's more to feast our eyes on around every bend. The sea breeze lashes out at the car, and deer, sheep and Highland cattle cross the road unannounced. So you don't get to daydream for too long. Fortunately, the combination of right-hand drive, driving on the left, swerving and changing gear keeps me alert. Around us, the orange-green appears in many guises. From tender moss beside mountain lakes to the Munros, as the Scottish mountains over 915 metres (3,000 feet) high are called. 'Munroing' has been an official sport here since Sir Hugh Munro mapped these peaks in the late 19th century. The challenge lies not in reaching the summit of the modest peaks, but in conquering the rough terrain. Wild swimming is another popular outdoor sport, and often mentioned in the same breath as Munroing. Although we choose the warmer of the two, we still explore the water from a packraft. The days seem to glide on by. In the evening the breathtaking camping spots throw themselves at our feet, and we plonk ourselves down around the fire with a hunk of bread and a dram of Scotch. For days our clothes are filled with the smoke that chased away the midges which otherwise would have eaten us alive.

The most northerly stop on our drive is my favourite: Sandwood Bay, Scotland's most inaccessible beach. We camp on the beach of Sheigra, a little village on a cliff, and leave as soon as the fog lifts. The fact that we almost can't find the trailhead is a good sign. If you get there early enough, you won't see anyone, even in high season. The rising sun plays with the colours of the landscape. The path is narrow and winds sharply uphill, before we're treated to a first glimpse of this slice of paradise. A white

Hatching a plan. *(left)* **Beach lookout near Sandwood Bay.** *(top right)* **Wild camping, deluxe.** *(bottom left)* **In my happy place.**

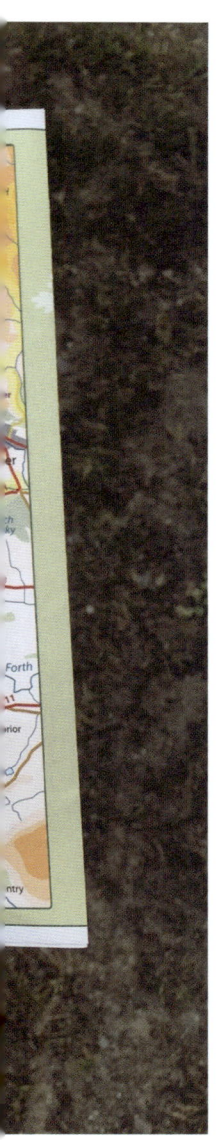

Scenic route into the mountains. *(left)* Campfire bonding. *(bottom left)* Old-school adventurers cross paths. *(top right)* Pack rafting on a loch. *(right)*

strip that is 1.5km (1 mile) long at high tide and neatly sheltered from the rest of the world by steep rock walls on both sides. You'll rarely find a more pristine beach. On a clear day you can even see the Cape Wrath lighthouse, which stands on the most north-western tip of Scotland. These cliffs are also teeming with life. From seals, sheep and hundreds of seabirds to the rare yellow bumblebee on one of the 200 plant species in the wild grassland.

Progress is being made, but there's still a long way to go. Today Scotland's Affric Highlands is already one of Rewilding Europe's nine official projects. More than 75% of the Scottish population voted in favour of the proposal to make it a Rewilding Nation. Making Scotland wild again, one tree at a time. //

In the evening the breathtaking camping spots throw themselves at our feet, and we plonk ourselves down around the fire with a hunk of bread and a dram of Scotch.

Overnight at Glen Affric, one of Europe's most successful rewilding projects.

LOCATION

Highlands
Scotland, United Kingdom

CURIOSITY

Following centuries of deforestation, the UK's northernmost country has invested heavily in rewilding. Now Scotland is poised to become the world's first 'Rewilding Nation'.

ADVENTURE TIME

The best times to visit Scotland are during spring – late March to May – and autumn – September to November.

TRAVEL TIP

Plan a trip along the North Coast 500, bringing together a route of just over 800 kilometres (500 miles) of stunning coastal scenery in the far north of Scotland. Wild camping spots and self-guided hikes abound.

BEFORE YOU GO

Think big, act wild and visit Rewildingbritain.org.uk to uncover the Rewilding Network and explore, start and support rewilding efforts.

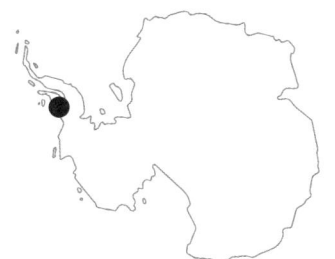

ANTARCTICA
69.9532° S, 75.1231° W

The Deep South

Crossing the Antarctic Circle on an expedition ship

Did you know that dinosaurs once roamed an ice-free Antarctica? The area around the South Pole is full of mysteries. For centuries, it has captured the imagination of sailors and storytellers alike. Here, unknown animal species survive alongside conspiracy theories about Nazis, aliens and Atlantis. Even the ancient Greeks spoke of a mysterious southern continent long before its discovery.

I too fall prey to its allure. I'm joining the crew of an expedition ship, in search of new climate narratives in Hope Bay and support on Pourquoi Pas Island, off the west coast of the Atlantic Peninsula. Its unblemished character has rarely been under so much pressure. We were supposed to be listening to whales, but heard something completely different instead: the sound of the climate crisis in action.

'Waves up to 8 metres [26 feet]... Looks like we're in for a Drake shake!' As soon as he sets sail, Captain Garcia can barely contain his excitement. From Punta Arenas we'll sail in only one direction for the next few days – south, via the Tierra del Fuego, out of the Strait of Magellan and past the legendary Cape Horn. We leave the protection of dry land behind us, to go out into the high seas, where the Pacific meets the Atlantic. A thousand-kilometre (620-mile) stretch of storm-prone open water separates the brave from the faint-hearted. Devoted to peace and science, Earth's last frontier is cut off from civilization by a ruthless natural border. During the heyday of polar exploration, the Drake Passage was an opponent to be feared. It was discovered as early as 1578, but most of the wooden sailing ships didn't survive the crossing. A rite of passage that I also can't avoid. Fittingly, the gateway to Antarctica remains commanding and unpredictable.

The ship shakes, rattles and rolls over the giant swell, in a battle against the wind and my vestibular system. In turbulent waters we're presented with procedures and taken through the most important safety drills: how to survive in polar temperatures. I try on a polar survival suit, which should provide up to 24 hours of protection in extreme conditions. The suit creates a microhabitat in which you can eat, drink and sleep, protected from waves, wind and rain. In the shaking belly of the ship, it feels especially claustrophobic. Meanwhile, I've turned a shade of green, but it's all par for the course. After 48 hours of battling against the wind, we sail into No Man's Land, through the Matha Strait. We've reached Antarctica. 'Land ice in sight!' shouts a scientist on deck, in keeping with tradition. We enter the frozen desert, leaving almost all other ships in the Southern Ocean behind. Beyond the Antarctic Circle at latitude 66°33'48.8" south, the temperature plunges noticeably. Starboard, the first iceberg appears. A welcome sign that may be up to 300 metres (100 feet) high, in shades of blue, white and grey. Colours that make your heart sing. It's instantly humbling, coming face to face with these majestic giants.

Touching ground
The long polar days take us via Pourquoi Pas Island deep into Hanusse Bay. On the way we pass one of the British bases, where the bright orange Sir David Attenborough polar research ship is anchored, having just completed its maiden trip. All around us it is quiet, muffled. There's nothing but the

First steps on the ice sheet around the continent. An almost sacred experience.

219

sound of the snow crunching against the bow. A few leopard seals shuffle across the ice. A snow petrel skims over the deck. We navigate through packs of snow ice several metres thick. It's also called sticky ice because the ice seems to surround the ship and stick to it, immediately making the bow heavier. We avoid the fast ice – land ice still attached to the continent – to keep the plates intact. The captain toils away, sometimes zigzagging, performing star-shaped manoeuvres so we don't get stuck. Behind us, the ice closes again as quickly as it opened up in front of us. Someone spots the fin of an orca feasting on krill that's been pushed to the surface of the open water. I don't know where to look first.

We sail right through the Gullet, a narrow channel full of ice blocks, surrounded by mountain peaks 2,700 metres (8,858 feet) high. The experience is raw and sacred at the same time. Once through the channel, the ship glides onto the ice bank of Hanusse Bay. Almost silently, like slicing through butter, the captain expertly places the ship in the snow so we can spend a polar night in this winter wonderland. We're in unknown, and therefore unreliable, territory. The powdery snow looks lovely in this light, but can camouflage breathing holes or cracks that run deep into the ice. The walkway is lowered and an expedition team goes on reconnaissance. Mesmerized, I take my first steps out into the snow, as if on another planet. This is a world where we're insignificant, a world without any traces of mankind. Antarctica wasn't dubbed 'White Mars' for nothing: astronauts are sent here to get used to the extreme conditions and otherworldly isolation. Here, too, the only temporary residents are scientists and researchers. An alienating and liberating feeling.

I explore the area around the ship, whose bow juts out a good 10 metres (33 feet) above the surface. An icebreaker, unlike ordinary ships, has a very special shape. You can recognize one by its characteristic 'nose' that lies on top of the ice, so it can use its own weight to break the ice below while sailing. The device dangling just above this is a seasonal sea ice measurement system, SIMS for short, which measures the

thickness of the ice. That data is transmitted via satellite to Bremen University. And samples are also taken manually: a bit further up, a scientist pushes, twists and drills holes into the ice. He is studying and measuring the thickness of the various ice layers. The ice bank we're moving on consists of 73cm (29 inches) of snow, 24cm (9.5 inches) of slush and 154cm (60 inches) of ice. A mere slice, given that some ice sheets here are up to 5km (3 miles) thick.

On the other side of the ship, another biologist launches a hydrophone into the water. Some inquisitive chinstrap penguins come to tune into the underwater sounds that the device amplifies. The recordings will later be sent off to marine biologists in the hope of providing them with new data on deep-sea life so far south. The world below the ice sheets is teeming with life; we hear seals chatting and penguins diving, as well as the otherworldly groans of crabeater seals. Up is down. Night remains day. Time is really relative here: theoretically, the South Pole is where all time zones meet.

Exploration trip

We set off for Charcot Island, a place that nobody has set foot on since it was discovered. As we're going to the deep south below the Antarctic Circle – and even below the symbolic 70th parallel south – there are no reliable maps of this area. From here on we enter uncharted territory that has only been captured by satellite images. And there is only one story we can base ourselves on. The logbook of Jean-Baptiste Charcot, who discovered the island in 1910 in his wooden sailing boat. I try to imagine how Charcot and his crew must have experienced this place, in such a small boat, with such limited resources. Madness.

A century later, two pilots are preparing a helicopter for take-off from the deck. Our expedition leader is going on a reconnaissance flight with the captain. Having hauled myself into a polar helicopter suit, I too am granted permission to go

A new perspective in Paradise Bay.

'We all have our own white south.
The continent exists as an excellent
metaphor for our personal journeys,
contextualizing the challenges of
our human nature.'

Ernest Shackleton

Passage through the Gullet,
a narrow channel full of ice
blocks, surrounded by
mountain peaks of up to
2,700 metres (8,860 feet).
(right) A full halo on
midsummer night.
It feels like it's me who's
beaming the most.
(following pages)

We skim over snow-capped peaks and ice cliffs 20 metres (65 feet) high. I get a taste of what it's like to be an explorer, if only briefly.

aboard as a silent stowaway. There'll be time to sleep later. We take off and explore a deep southern area where no helicopter has ever flown before. We skim over snow-capped peaks and ice cliffs 20 metres (65 feet) high. I get a taste of what it's like to be an explorer, if only briefly.

Once we're back on the ship, we head out for a longer trip into the snow. It's hard to tell where the sky ends and the ground starts. The light is blinding. We step into a disorientating vacuum, in silence, everyone in one line. The fins in the ice are snow-covered pieces of pack ice that were blown ashore and that could flip over under our weight. Distances and heights are difficult to estimate. Behind us, the ship floats in a horizonless white plain, as if sailing through the clouds. In front of us, a few crabeater seals are sleeping. Yesterday we spotted the rare Ross seal, which had the scientists shedding tears of joy and dancing on the deck. When the midnight sun manages to push through the clouds, it treats us to a full-blown halo, a light phenomenon that takes place in high, thin clouds made up of ice crystals. Everything is superlative here. And it feels like it's me who's beaming the most.

I think again about the cosmos. Many astronauts who view our planet from space experience a cognitive shift called the 'overview effect'. 'It does something to you', said American astronaut Edgar Mitchell when he returned from his Apollo 14 mission in 1971. Something great had happened to him up there, an overwhelming realization of the interconnectedness of all life. Since the dawn of human space travel in the 60s, people have come back from their missions with the feeling that they have experienced something 'up

Landing on the helideck after an initial ice reconnaissance. *(top)* **Meeting a group of curious chinstrap penguins.** *(bottom)* **Nowhere I'd rather be.** *(right)*

there' that we need down here. They describe that 'something' as an overwhelming love, as admiration or a sense of being in an epic. Many astronauts return with a missionary urge, as ambassadors for sustainability and nature conservation. In the blink of an eye, these astronauts understood that we are connected, not just to other humans, but to the planet itself. I doubt I'll ever get any closer to that feeling than I do here.

Hope Bay

At the very tip of the rocky coast of Paradise Bay, we find the first sign of human life: the Argentine base of Almirante Brown. The scientific research centre is one of 13 small Argentine bases, and one of 68 international bases on the continent. Initially a permanent base, it is now only open during the summer season. However remote this place may be, it is bursting with light and life. Around the base waddles a colony of Adélie penguins. In the rock cavities near the base

Towering icebergs gather in the bay in the form of ice-cream cakes, triumphal arches and crystals, in a watercolour palette of blue and white.

we find a rare *Belgica antarctica*, a species of flightless midge. Small yet plucky, it is the only insect on the continent – and in winter, the largest land animal, roughly 3 millimetres (a tenth of an inch) in size. Antarctica is a soft desert for those that can adapt.

We climb higher up the mountain and the continent, gazing at giant ridges on the horizon. Sloshing about, with snow up to our knees as it snowed much more than usual this year. We balance on a rim a good 3 metres (10 feet) thick. Along the coastline, the colony waits in vain for the rocks to emerge from under the snow, so that they can once again lay their eggs there this year. As we observe them, we're regularly startled by the sound of cracking ice. More writing on the wall. Towering icebergs gather in the bay in the form of

ice-cream cakes, triumphal arches and crystals, in a watercolour palette of blue and white. These giants are made of land ice, fresh water that breaks away from the continent and floats into the ocean. Because it contains no salt, it is rock hard. It was a hunk of ice like this that previously served as proof that the continent existed. The mountain groans until the ice slips into the bay. Air bubbles that have been trapped there for centuries escape through cracks and crevices. A soundtrack with a final chord that's been a long time coming. A sad ode to deep time.

While in maps on many school walls this wilderness is reduced to a white stripe, or

Behind us, the ice closes again as quickly as it opened in front of us. *(left)* **The rare Ross seal, on the fragile sheet.** *(top)* **Ice formations in a muffled landscape.** *(bottom)*

sometimes doesn't even make it onto the map, it contains more than 90% of all the ice mass on Earth. Yet this canary in the climate-crisis coal mine is also what gives it the most hope. WWF has managed to ban fishing from an ocean area of 4,500 km² (1,740 square miles) off the northern Antarctic Peninsula, naming the wildlife sanctuary Hope Bay. The bay is teeming with natural intelligence and all creatures great and small, including whales, krill and phytoplankton. We almost forget that the deep ocean and its inhabitants are our natural carbon processors. Each year, marine ecosystems and wildlife together absorb nearly three billion tonnes of carbon. The amount of CO2 a whale absorbs is comparable to the amount absorbed by around 3,000 trees. This blue carbon phenomenon captures more carbon worldwide than all rainforests combined. And that is more than worth protecting.

This is the only place on Earth that doesn't officially belong to anyone. For more than 4.5 billion years, no human was carried here on the wind or the waves. The explorers who first mapped the continent were hailed as heroes, but carnage often followed in their wake. The tide turned in 1959, and the continent became the first open nuclear-free zone following the Cold War arms race. My homeland Belgium was one of the 12 countries that co-signed the Antarctic Treaty. The Protocol on Environmental Protection was later added, making it impossible to exploit mineral resources until at least 2048.

It is our most symbolic place, dedicated to peace and science. But that doesn't mean there aren't nations that have unofficially claimed a chunk of the Antarctic. Officially, the 68 stations are solely engaged in peaceful scientific research, but their motivations are not always so clear. Now it's up to us to protect this wilderness and its inhabitants. For activists, restoring the whale population is a much-needed reminder that international cooperation can be effective, protection can work, and our species has the ability to reverse the damage we've done to our natural world. So long as there's a whale plunging its tail into the depths, there's hope on the horizon. //

Abbey Road, Antarctic edition.

Adelie penguins near the Almirante Brown polar station. *(left)* Scientists take samples and measure the ice layers. *(bottom left)* Paradise Bay. *(right)* A leopard seal carries on sleeping undisturbed on an iceberg. *(bottom right)*

LOCATION

Polar circle
Antarctica

CURIOSITY

Antarctica is the highest, driest, coldest and windiest continent on Earth, and doesn't belong to anyone. It is a white desert full of life, containing most of the world's fresh water.

ADVENTURE TIME

During the summer period, from November to March, wildlife is most active and the days are at their longest. In winter, sea ice envelops the continent and Antarctica is plunged into months of darkness, with temperatures to −60 °C.

TRAVEL TIP

Visiting Antarctica is a privilege and a responsibility at the same time.

BEFORE YOU GO

Watch the ecologist documentary adventure that is *Sanctuary*, in which Álvaro Longoria and brothers Javier and Carlos Bardem launch themselves into a Greenpeace campaign in the hope of securing protected status for a vast area of the Antarctic Ocean.

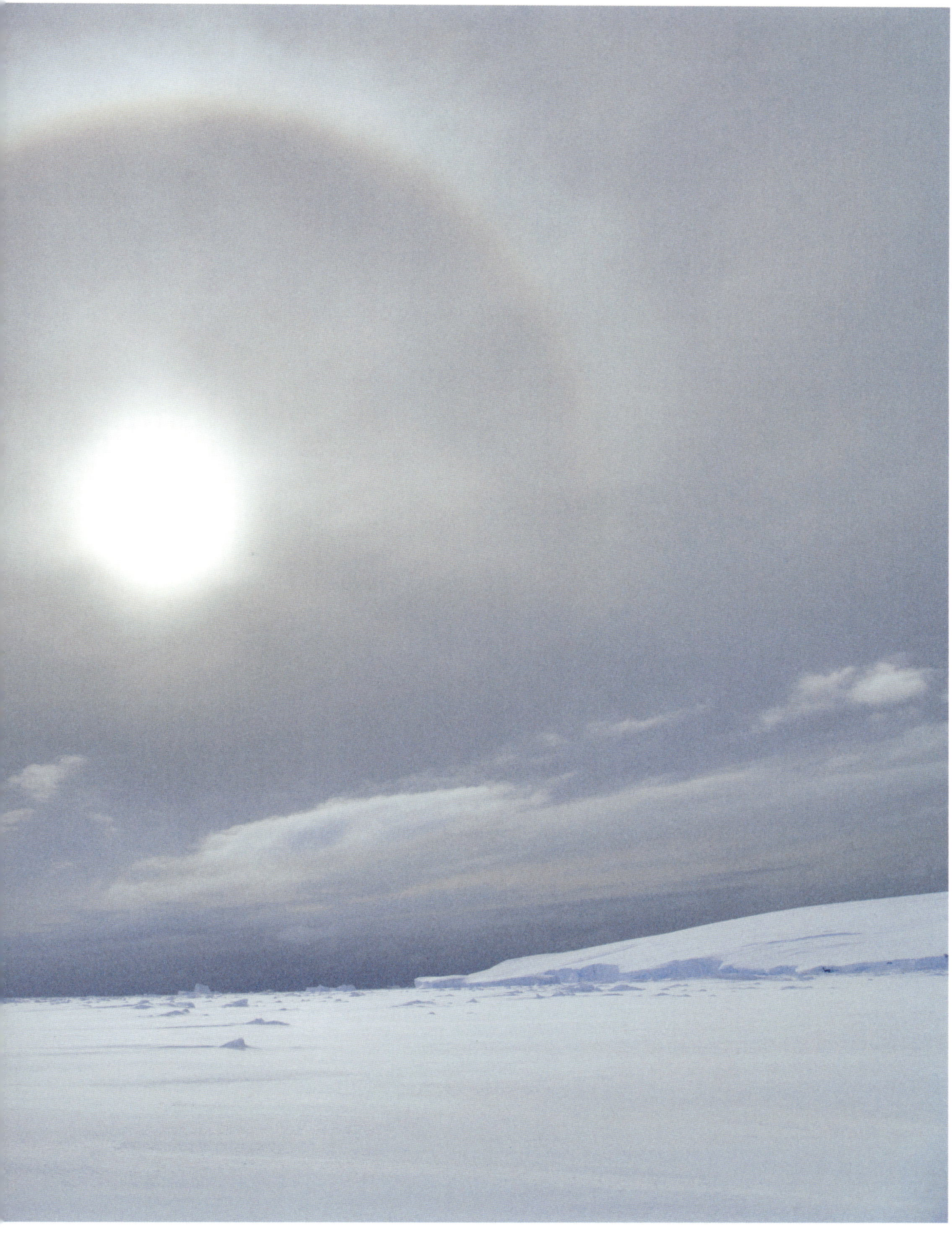

NAVIGATE WITH FRESH EYES

10 essentials for 21st-century explorers

1. Let go. Open up.
Adventure travel is about exploring the possibilities and being surprised. Embrace the uncertainty. Look at the world through the eyes of a child. Stop planning, just let it happen.

2. Pack light. Do more with (much) less.
Everything you take with you will weigh you down. Not just physically, but also mentally. Be strict. Take only what you really need, or can be given away.

3. Just get going.
There'll always be a reason to delay. Whether you've got a week or six months, set yourself a goal and leave. It doesn't need to be far or expensive. Put one foot in front of the other, and go.

4. Be active, not proactive.
Nothing is more rewarding and sustainable than travelling under your own steam. Earn your rest day and sweat yourself into a groove. Strengthen your senses. Get bored. Live.

5. Believe in goodness. Be resilient.
Fear is a bad adviser. Believe in the goodness of people. Start from the positive, but be prepared for the worst-case scenario. Bumps in the road make for the richest encounters.

6. Immerse yourself.
Mix with the locals. Buy most of your food and other items at local markets. Sleep in people's homes. Read a book by a local author. Don't plan your days to the max. Instead, live like the locals.

7. Stay honest. Be truthful.
What do you believe in? Which principles do you stand up for? Live by your deepest values. Forget 'likes' and the curated world of the internet. *Esse quam videri*: 'to be, rather than to seem.'

8. Know your limits. Follow your intuition.
Something doesn't feel right. Too risky or difficult? Listen to your gut, not peer pressure. Don't underestimate the forces of nature. Sometimes it's braver to stop.

9. Be critical. Feel the tension.
Even the best stories have a flip side. Think twice before taking part in anything or handing over money. Watch out for exploitation, animal cruelty and destruction. Don't shy away from that tension. Call it out. Share your doubt.

10. Be responsible. Be an ambassador.
Take care of yourself and your host country. Behave respectfully towards the local people and creatures. Don't leave anything behind. Our planet and its ecosystems are the most precious treasure we have. Inspire others to actively protect our biodiversity.

'And it was always the stories that
needed the telling that gave us the rope
we could cross any river with.
They balanced us high above any
crevasse.

They made us be natural acrobats.
They made us brave. They met us well.

They changed us.
It was in their nature to.'

Ali Smith

ON THE SIDELINES

Through the looking glass

Adventures give you the time to explore, get bored and be reborn. They help you try to shed your old skin and come back different. Suddenly the hours pass by more slowly. You allow yourself to stare into space for an hour, to watch a rainstorm pass over. You spend six hours on a rock in no man's land waiting for the border to open, or not. What is classed as unproductive at home turns out to be more than worth the wait elsewhere.

With little more than your travel outfit on your back and your passport in your pocket, you quickly peel yourself back to the essentials. The further you move away from your own comfort zone, the more fragile the roles, patterns and routines you thought defined you seem to become.

While on the road I was given a warm send-off by grandmothers, sons, stepfathers and uncles. I shared tables with bikers, carpenters, cleaners, shepherds and soldiers. I was offered a bed by people of all walks of life. Team Human in all its glory, shaped by its own belief in what is good (for us). That picture is rarely black and white, but often not rainbow-coloured either.

Alter ego

In 14 out of 20 countries and 13 out of 16 states, I travelled with a made-up backstory. Along the way I was forced to dive back into the closet, and as a woman I really had to stand my ground. In the weeks that gave me the greatest feeling of freedom, it turned out to be unwise to be completely myself. Travel buddies or friends in photos were promoted to my husband, and I made up various storylines, which I often had a really good laugh about. But inside it started to grate, not being able to be honest everywhere I went on those journeys that brought me closer to myself. Not to be addressed directly, as I was the only woman in the group. I have rarely felt so warmly welcomed and unwelcome at the same time. I've never played hide-and-seek quite so liberally.

True colours

Often you only meet yourself on your return. Like the 'Welcome Home' sign, you seem to consist only of a collection of independently fluttering elements. Once familiar roles and environments no longer keep you together, you have to get started again with a DIY kit. Astronauts also suffer from a slight orientation problem and growing pains after a period of prolonged weightlessness.

Would they still help me on my way if they knew who I really was? Would I be safe? The fact that I kept asking myself those questions – and double-checking all my public posts during an unplanned overnight stay at the headquarters of a radical right-wing biker gang – gave even the nicest of stories a sad undertone. It shouldn't still be like this.

Being out and proud was for a long time not for me. Too loud, too conspicuous, too activist, too... Until I finally felt that this no longer had to be about me. Ultimately, it is a sense of togetherness that prevails. I'm not putting forward my sexuality, but everyone's freedom. Being loud and clear for those who don't have their own voice. My love knows no bounds anymore. //

A world map according to fish,
a Spilhausprojection. In
1942 Athelstan Spilhaus
created the first ocean
centered map of the world.
This projection centralizes
Antarctica and stresses the
connection between all
oceanic basins.

THANK YOU

To our interconnected world, plankton, wild forests and the ocean.

To the travel and life partners who accompanied me along the way.

To all the generous strangers who opened their homes and hearts to me.

To Trui, Gaea, and all the women who showed me what adventure is.

To the humble whose often invisible choices are healing the world.

To Carolijn, Sarah and Lannoo Publishers for their trust in me.

To my granny, the sweetest non-traveller I know.

TEXTS

Lien De Ruyck

COPY EDITING

Léa Teuscher

TRANSLATION

Heather Mills

IMAGE SELECTION

Lien De Ruyck

PHOTOGRAPHY

All images are © Lien De Ruyck, except for:

p. 16, 19, 20, 22, 24	© Andrew Phelps en Paul Kranzler
p. 14, 44, 46, 48	© Alexander Dumarey
p. 58, 61, 63	© Sander De Wilde
p. 54, 55	© Marjolein van Roosmalen
p. 56	© Hans Jellema
p. 84	© Aston Shannon
p. 98	© Outdoor Animation
p. 28, 108, 110, 112, 115, 116, 118, 175, 177, 182	© Ine Dehandschutter
p. 191, 192, 193, 200, 201, 203, 204, 205, 206	© Katleen Willaert
p. 194	© Jochen Verghote
p. 100, 106, 221, 222, 223, 224, 227, 228, 229, 231, 232-33	© Mike Louagie

BOOK DESIGN

Carolina Amell

TYPESETTING

Han van de Ven

If you have any questions or comments about the material in this book, please do not hesitate to contact our editorial team: art@lannoo.com

© Lannoo Publishers, Belgium, 2022
D/2022/45/241 - NUR 450/500
ISBN 9789401485845